Climate Change in Practice

Climate science shapes important decisions about the structure of economies, human development, natural resource use, and ways to reduce vulnerability to extreme weather. Derived from an undergraduate course on Climate and Society taught by the author, this accessible, full-colour, book seeks to challenge and provoke readers by posing a series of topical questions concerning climate change and society.

Topic summaries provide answers to technical, socio-economic and moral questions surrounding the deployment of climate science. These include: how to build and test a climate model; who or what is most at risk from climate change; how to decarbonise economies; how it is possible to adapt to uncertain climate; and whether we should geoengineer the climate. Practical exercises and case studies provide deeper insights by taking readers through role-play activities and authentic climate change projects. These also enable development of transferable skills in data evaluation, and resource and risk assessment. The book is supplemented by various supporting materials including notes for instructors and students, graphics, video clips, games and online resources available at www.cambridge.org/wilby, offering scope for further private study and group work. With nearly 600 references, reflecting seminal work and the most recent research in the topic areas, this book is ideal for students of geography, natural science, engineering and economics, as well as practitioners involved in the climate service industry.

ROBERT L. WILBY is Professor of Hydroclimatic Modelling in the Department of Geography at Loughborough University, UK. His research expertise covers climate risk assessment and adaptation planning for freshwater systems, regional climate downscaling and hydrological modelling, climate services, capacity development, and environmental monitoring and management. In addition to his academic positions, Professor Wilby has spent time at the US National Center for Atmospheric Research, as well as in commerce for Severn Trent Water, in consultancy for the Department for International Development, the World Bank and WWF, and in government at the Environment Agency.

'The most engaging book I have read about climate change in many years. I found it difficult to put down! It takes a fresh and novel perspective on climate change science, through critical analysis of the utility and applicability of the science for use in society. Comprehensively researched and up to date, many of the issues discussed relate to recent debates and developments in the field of climate change science and policy. At each stage, the scientific and societal challenges of the production and use of the scientific material is considered. The exercises are novel, thought-provoking, and will engage today's students – I look forward to applying them in my own classroom. The book is extremely well written, in accessible language, and is brimming over with relevant and attractive photographs, maps and diagrams. It will appeal to students, instructors, decision-makers, and researchers of climate change alike.'

Rachel Warren, University of East Anglia

'*Climate Change in Practice* is an excellent and much needed book that challenges the reader to think broadly about a range of issues surrounding the climate change debate. The author has a unique range of expertise and experiences to tackle such broad topics. While intended for students and practitioners in the climate services industry, the book also provides a very valuable context not only for the climate-interested public but also for climate researchers, regarding the technical, socio-economic and moral questions surrounding the applications of climate science.'

Judith Curry, Georgia Institute of Technology

'Wilby's *Climate Change in Practice* could hardly have come at a more timely moment as the debate around climate change mitigation is sharpened by recent political developments. Illustrated throughout by relevant real world problems, accessible to the lay person, and informative for experts, this is an enjoyable and fascinating read for anyone with an interest in one of the major challenges for human society in the 21st century. Addressing the climate change science-policy interface, it provides a discussion of competing perspectives based on critical evaluation of supporting data and analyses. Wilby is uniquely well qualified to write about this boundary area, with a background that includes academic research, a senior advisory role for government, and extensive consultancy experience, and the book consistently impresses with its wide range of authoritative and up-to-date material.'

Howard Wheater, University of Saskatchewan

Climate Change in Practice

Topics for Discussion with Group Exercises

ROBERT L. WILBY
Loughborough University

CAMBRIDGE
UNIVERSITY PRESS

CAMBRIDGE
UNIVERSITY PRESS

University Printing House, Cambridge CB2 8BS, United Kingdom

One Liberty Plaza, 20th Floor, New York, NY 10006, USA

477 Williamstown Road, Port Melbourne, VIC 3207, Australia

314-321, 3rd Floor, Plot 3, Splendor Forum, Jasola District Centre, New Delhi - 110025, India

79 Anson Road, #06-04/06, Singapore 079906

Cambridge University Press is part of the University of Cambridge.

It furthers the University's mission by disseminating knowledge in the pursuit of education, learning and research at the highest international levels of excellence.

www.cambridge.org
Information on this title: www.cambridge.org/9781316507773
10.1017/9781316534588

First published 2017

A catalogue record for this publication is available from the British Library

Library of Congress Cataloging in Publication data
Names: Wilby, R. L. (Robert L.)
Title: Climate change in practice : topics for discussion with group exercises / Robert L. Wilby, University of Loughborough.
Description: Cambridge : Cambridge University Press, 2017. | Includes bibliographical references and index.
Identifiers: LCCN 2016037638 | ISBN 9781107143456 (hardback) | ISBN 9781316507773 (pbk.)
Subjects: LCSH: Climatic changes–Study and teaching–Activity programs. | Climate change mitigation–Study and teaching–Activity programs. | Environmental education–Activity programs.
Classification: LCC QC903 .W567 2017 | DDC 363.738/74–dc23
LC record available at https://lccn.loc.gov/2016037638

ISBN 978-1-107-14345-6 Hardback
ISBN 978-1-316-50777-3 Paperback

Additional resources for this publication at www.cambridge.org/wilby.

To Dawn and Samuel
Love finds admission, where proud science fails.
Edward Young, *Night Thoughts* (1742–1745)

Contents

Preface

This book has been road-tested by students for students. The Climate and Society module at Loughborough University has been running since 2009 and attracts up to 70 undergraduates each year. Students are drawn from human and physical branches of the geographic tradition, as well as from allied courses in economics and management. Trans-disciplinary perspectives are integral to the module content. Such is the enormity of the issues and breadth of science involved that climate change cannot be fully comprehended by individual researchers or disciplines. This leaves plenty of space for ambiguity or ignorance about unresolved issues.

This book is designed to challenge and provoke by posing topical questions concerning climate change and society. For instance, how do we measure the temperature of the Earth, or define dangerous climate change, or even decide who is a climate expert? Anthropogenic climate change presents huge moral and ethical dilemmas about who gets protected as well as whether there should be limits to economic and technical fixes. There are divergent opinions about the trustworthiness of climate models and the extent to which climate services can really build resilience for the vulnerable. Climate communication can be framed and manipulated in so many ways. Such topics are ripe for group discussion as well as individual learning.

Each chapter follows the same pattern. To begin, there is a topic summary with suggestions for background reading. These lead into a short essay on the question posed. Again, the intention is to lay out key issues and concepts rather than to attempt an exhaustive critique. This is followed by a group exercise to reinforce points. Finally, learning outcomes, questions and further resources are provided for personal study. Answers and notes for each exercise are provided online. The novelty of the book lies in the dual approach of strengthening technical knowledge and practice for individuals, whilst exploiting opportunities for learning and debate within large groups.

The book is not intended to be a core text but is best regarded as a supplementary handbook. Some prerequisite awareness is assumed and this is readily acquired from *Why We Disagree About Climate Change* (Hulme, 2009), *The Thinking Person's Guide to Climate Change* (Henson, 2014), or the *Routledge Handbook of Climate Change and Society* (Lever-Tracy, 2010). Each captures the essential concepts and development of climate change as a physical, social, political, cultural and technical phenomenon. However, *Climate Change in Practice* extends these perspectives into distinctly applied realms. For example, one game explores the utility of seasonal climate forecasts in hazard management. Other role-play activities shed light on climate negotiations, or the practical steps involved in a climate risk assessment.

Climate Change in Practice provides material to support about 40 hours of contact time comprising core content on the workings of the climate system, regional consequences of climate change, and societal responses. Running in parallel are the practical exercises and group work to embed and apply knowledge. The book is also intended to support modules in environmental science and engineering, applied climatology, development studies, natural resource management, economics, management and planning.

Target audiences are university students and their instructors, postgraduate students and researchers, as well as climate service providers. Each chapter provides a ready-to-go topic and exercise that can be customised by tutors to better meet their needs. Topics can stand alone, so there is also the option to draw on content more selectively (although the wide-ranging subject matter is inherently integrated). Researchers and students will discover entry points to some very important and unsettled scientific challenges, with plenty of avenues for deeper enquiry. Chapter questions may be helpful for revision or pinpointing gaps in understanding.

Topic primers supplement core material by presenting different viewpoints and a more coherent understanding. For example, climate models could be defined only in terms of their physical components, but this would not begin to unpack their wider societal relevance and policy impact, or the different ways in which climate model information can be evaluated, used and abused. Likewise, geoengineering could be framed as a purely technical endeavour, but this would overlook the immense moral and ethical concerns circling these nascent technologies.

The content is intended to be applicable regardless of region or curriculum. However, special attention is given to the climate vulnerabilities of developing regions and built environment. Such emphases are justified given the high exposure to climate hazards and adaptation deficit of these communities. For instance, climate threats to rapidly expanding megacities in coastal zones of Africa and Asia are of particular concern. Water-related risks and opportunities figure prominently too because this sector is so central to the future evolution of climate impacts and societal responses.

The topics reflect the research interests and technical advisory work of the author. These spring from more than two decades of experience of translating climate science into practice through projects in Australasia, Central and South Asia, Europe, North America, the Middle East and North Africa. Case studies include heat and air quality management for London, food and water security for remote villages in Yemen, hydropower planning in Tajikistan, climate risk assessment for Djibouti, and water resource management in the Colorado Rockies. All are based on first-hand experience of these very diverse projects.

The author is not a social scientist but has tried to draw on some of the helpful perspectives of this discipline, especially in later chapters on climate communication, ethics and scepticism. By the end of the book it is hoped that readers will recognise that *Climate Change in Practice* brings together contested political, scientific, economic and moral views about priorities. Even so, much can still be accomplished through pragmatic and wise application of climate science. Prospective climate service providers should also recognise that *how* we choose to deploy our knowledge is just as important as *what* we know.

To conclude, it is vital that climate change be viewed alongside other major challenges ahead. Not least amongst these are concerns about poverty, demographic and technological change, pandemic, resource exhaustion, loss of biodiversity, and regional conflict. Hence, dogmatism about climate change as a singular threat is unwise. It is far smarter to regard climate change as connected to these other challenges, and to use such insight to find more integrated solutions.

Acknowledgements

This work brings together many years of climate research and practice. My thinking and experiences have been shaped by time spent alongside colleagues at the University of Derby, King's College London, Lancaster University, the Environment Agency, the Department for International Development, the European Bank for Reconstruction and Development, Acclimatise, and the World Bank.

This career path has weaved between roles in academia, consulting and government, but friends and colleagues at the Department of Geography at Loughborough University have been a constant source of encouragement and inspiration. I hope that their collegiality and respect for different disciplines shines through my work. Likewise, my students have provided constructive feedback and given me immense satisfaction through their achievements.

I am particularly indebted to the following people. Greg O'Hare has mentored and tested my writing along the way. His wit and incisive comments have often caused me to stop and think more deeply. Over the years, there has been much fruitful collaboration and maturation of my research thanks to Tom Wigley, Linda Mearns, Kathy Miller, David Yates and Martin Clark at the National Center for Atmospheric Research in Boulder. Projects with Conor Murphy and his team at Maynooth University are always a tonic and reminder of the joy that comes from scientific inquiry. Christel Prudhomme at the Centre for Ecology and Hydrology has been a source of much inspiration.

There are also those who have challenged my assumptions or brought a fresh perspective through their work. Suraje Dessai rightfully speaks to us of the importance of the local context in shaping climate vulnerability; Bruce Hewitson of the need for integrity in climate services; and Hayley Fowler of the virtue of more process insight from regional climate modelling. Glen Watts and Harriet Orr set examples of how to bring climate science into policy and practice.

I would also like to take this opportunity to thank Susan Francis and Zoë Pruce at Cambridge University Press. They have offered plenty of helpful support and advice – sometimes in very diplomatic language! Their experience and guidance has been greatly appreciated.

Finally, friends and family have unknowingly contributed. Tony Chanter, Tim Daniel, Chris Dawson, Paul Hodgkins, Ed Kase, Richard Wolniewcz and Andrew Frankenburg have debated many big issues from a layman's perspective, whilst quaffing fine ale. Kitty, Philip and Derek have always kept their brother firmly grounded. Mum and Dad have given unerring support. Dawn and Samuel remind me why it all matters.

1

What Is the Global Mean Temperature and How Has It Changed?

TOPIC SUMMARY

Accurate measurement of the Earth's temperature is far from straightforward. Given the diversity of land and ocean environments, even the concept of a global average is hard to grasp. Yet the global mean surface temperature (GMST) record is an important reference point for climate modellers and policy-makers alike. In 2015, the GMST was about 1 °C above the average for 1850–1900, but the underlying data show considerable regional and decadal variability in rates of warming. Some claim that the rise in GMST slowed (or even halted) for about 16 years after the strong El Niño event of 1997/98. Explanations for this behaviour include statistical artefacts, data gaps and uncertainties, natural climate variability, weaker solar forcing, and increased heat sequestration in the depths of the Pacific and Atlantic oceans. Others resolutely deny the existence of a pause in warming and point to rising global ocean heat content (GOHC) as a clear indication of planetary heat gain over time.

BACKGROUND READING

Stephens et al. (2012) and von Schuckmann et al. (2016) explain how understanding of the Earth's energy imbalance is improving by combining top of atmosphere radiation (satellite sensors), surface air temperature (meteorological data) and ocean heat content (buoys) monitoring. Jones and Wigley (2010) critique the main sources of uncertainty in various global mean temperature records.

1.1 RISING TEMPERATURES, RISING CONCERNS
The Power of a Graph

The annual global mean surface temperature (GMST) record is an iconic but highly contested graphic (Figure 1.1). No one can 'feel' a global mean temperature, so it is also a rather abstract concept. Just as my annual car mileage gives a sense to which a year has been more or less than average, the same number of miles can be accrued in very different ways. Likewise, the GMST tells us nothing about the large regional variations in warming (and cooling), whether on land or over the ocean, that contributed to that global mean (Figure 1.2).

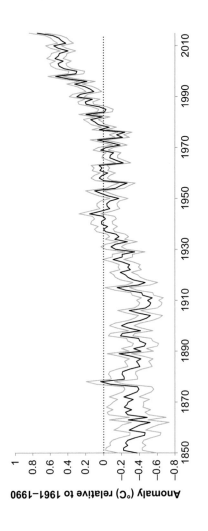

FIGURE 1.1 Annual global mean temperature anomalies (black line) based on analyses of surface data (HadCRUT4) showing the 95% confidence interval of the combined effects of measurement and sampling, bias and coverage uncertainties (grey lines). Annual anomalies are calculated with respect to the 1961–1990 mean. *Data source:* Jones et al. (2012) and UK Met Office www.metoffice.gov.uk/hadobs/hadcrut4/

Annual J-D 1997-2015   L-OTI(°C) Anomaly vs 1961-1990   0.51

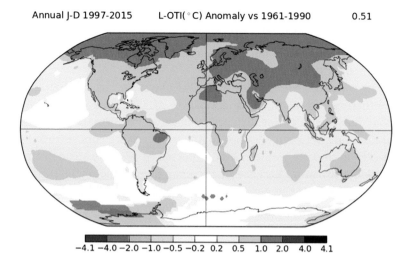

−4.1 −4.0 −2.0 −1.0 −0.5 −0.2 0.2 0.5 1.0 2.0 4.0 4.1

FIGURE 1.2 Annual mean surface temperature anomalies for the period 1997–2015 compared with 1961–1990. *Data source*: GHCN_GISS_ERSSTv4 http://data.giss.nasa.gov/gistemp/ [accessed 25/1/16]

The GMST does show that there has been an inexorable rise in temperatures of about 1 °C above the 1850–1900 average, with most rapid warming for 1910–1930 and 1980–2000. The simplicity of the plot belies the complexity of the underpinning data and analysis, which have attracted fierce debate (Chapter 4). This is because the GMST is a benchmark for testing models of climate change, and a reference point for policy-makers. Undermine the credibility of the global mean temperature record and you undermine one of the most important building blocks in climate science.

Global Warming Hiatus

In 2013 the Daily Mail[1] claimed that "*global warming has stalled since 1998*". Citing an apparent slowdown in GMST since the late 1990s (Figure 1.1), the tabloid went on to question the trustworthiness of climate models that had predicted higher rates of warming than observed. Their argument was appealing given that greenhouse gas

---

[1] www.dailymail.co.uk/sciencetech/article-2259012/Global-warming-Met-Office-releases-revised-global-temperature-predictions-showing-planet-NOT-rapidly-heating-up.html

(GHG) emissions continued to rise throughout the period. However, with new records set in 2014 and 2015, global warming seems to have resumed, with notable temperature rises observed in the Arctic and high latitudes (Figure 1.2).

The scientific community is divided about how to handle these conflicting messages. Lewandowsky et al. (2016) assert that the issue has been a major distraction because the climate is known to fluctuate between decades. In any event, rising concentrations of carbon dioxide would not be expected to produce steady rates of global warming because of the changing efficiency of carbon sinks. Moreover, the period 2000–2015 is simply too short to be representative of longer-term trends. Rajaratnam et al. (2015:ii) simply state that there is "*no hiatus in the increase in the global mean temperature, no statistically significant difference in trends, no stalling of the global mean temperature, and no change in year-to-year temperature increases.*" Cahill et al. (2015) add that there is no evidence of any detectable change in the warming trend since the 1970s. All three studies conclude that the terms 'pause' and 'hiatus' are not statistically consistent with the observed GMST record.

Room for Doubt

Others are more circumspect. For example, Hawkins (2015b) believes that the period provides an opportunity to establish whether the climate is evolving in line with scientific expect-ations. Curry (2014a) is convinced that observed temperatures contradict the prediction of the Intergovernmental Panel on Climate Change (IPCC) that warming would proceed at ~0.2 °C/decade at the beginning of the twenty-first century (the actual value is thought to be ~0.1 °C/decade). Curry further claims that "*the warming hiatus raises serious questions as to whether the climate model projections of 21st century have much utility for decision making, given uncertainties in climate sensitivity to carbon dioxide, future volcanic eruptions and solar activity, and the multidecadal and century scale oscillations in ocean circulation patterns.*"

However, it is important to recognise that the putative global warming pause begins with a strong El Niño event in 1997/98. This means that any linear trend in GMST estimated from that point onwards will be negative or at least neutral until much hotter years are encountered. [In fact, 1998 remained an outlier until it was surpassed in 2005, 2010, 2014, and then in 2015.] Despite there being only two weak (2004/5, 2006/7) and two moderate (2002/3, 2009/10) El Niños in the 12 years following the strong event of 1997/98, the period 2001–2010 was still the warmest decade on record. Hence, depending on how observed and modelled GMST data are statistically analysed, decades with no trend yet highest mean can be found (see Easterling and Wehner, 2009). These findings are not incompatible.

Global Temperature Monitoring Networks

Legitimate questions surround the quality and density of temperature measurement networks on which the GMST depends. One of the difficulties in checking the existence of the hiatus relates to a lack of temperature data for all parts of the climate system (i.e. land areas including high altitudes, upper atmosphere, areas covered by snow and ice, various depths in the ocean). Callendar (1938) was the first to attempt to create a global temperature record (Figure 1.1). His original study covered the period 1880 to 1930 and relied on less than 150 land stations, with the vast majority situated in the Northern Hemisphere. Although there are more stations today, their distribution is still biased towards northern, temperate regions (Figure 1.3). However, other temperature data are now compiled from satellite observations, ships and floats (see below).

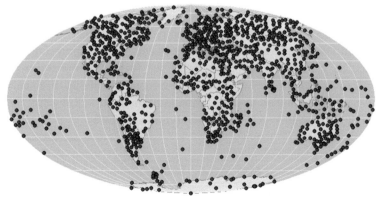

GCOS Secretariat, 1 March 2014

FIGURE 1.3 Global Climate Observing System (GCOS) surface network in 2014 (1017 stations). *Source*: World Meteorological Organisation

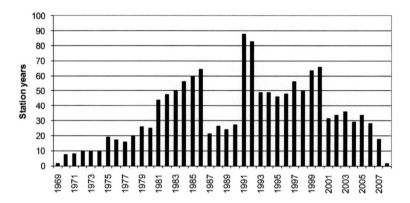

FIGURE 1.4 Amount of daily rainfall data held by the National Water Resource Authority, Yemen, for the period 1969–2008. The impact of periods of conflict in the late 1980s and post 2000 are evident. *Source*: Wilby and Yu (2013)

Year-to-year changes in the distribution and number of meteorological stations account for some of the uncertainty in the annual mean temperature anomalies shown in Figure 1.1. Confidence is greater now than in the nineteenth century because of improvements to the observing network. Nonetheless, successive reports on the status of the Global Climate Observing System (GCOS) Surface Network Monitoring cite 1017 meteorological stations in 2014 compared with 1040 in 2011 (Figure 1.3). The network shrank slightly and still has low densities/reporting frequencies for areas in sub-Saharan Africa, Central Asia and South America. More dramatic declines in national network density or data quality may be linked to times of civil unrest and/or wholescale changes in equipment (Figure 1.4). For instance, a marked drop in recorded sea surface temperatures in 1945 coincided with a shift from engine room intake measurements (on US ships) to uninsulated bucket measurements (on UK ships) at the end of the Second World War (Thompson et al., 2008).

Measurement and Station Errors

Using corrected and updated surface temperature data, Karl et al. (2015) claim that there was no discernible (statistical) decrease in the rate of global warming between the periods 1950–1999 (+0.113 °C/decade) and 1998–2014 (+0.106 °C/decade). Their analysis showed strong warming trends in the Arctic since 2000, but marked Northern Hemisphere mid-latitude cooling over the same period. However, Curry (2014a) is concerned that the techniques used by Karl et al. (2015) to fill data gaps over land, open water and changeable sea ice surfaces in the Arctic could have biased the analysis. Intriguingly, they did not incorporate best available ocean temperature data (from the Argo array, see below) so Curry (2015) concluded, *"uncertainties in global surface temperature anomalies [are still] substantially understated".*

Others are concerned about possible errors introduced by poor siting (Table 1.1) and maintenance of meteorological stations (Davey and Pielke, 2005). A major worry is that estimates of global mean temperatures for land areas are artificially raised by the influence of sites with urban heating. In practice, the site selection criteria used by GCOS are weighted against stations in heavily populated areas but in favour of rural locations. Where it is difficult to find rural sites, techniques may be applied to avoid, assess or compensate for possible urban heat island effects (Parker, 2010). For example, comparison of temperature trends under windy and calm weather can reveal urban warming, which is most likely during still, night-time conditions (Parker, 2004). Temperature records in China have attracted particular attention because of the astonishing pace of urban development. Jones et al. (2008) report that urban-related warming in China since 1951 is about 0.1 °C/decade compared with regional-scale warming of 0.81 °C over the same period. A global assessment of temperature trends at 39,028 weather stations found an average difference between urban and rural sites of less than 0.01 °C/century (compared with observed warming over land areas of 1.9 ± 0.1 °C/century since 1950) (Wickham et al., 2013).

Changing Solar Radiation

Some believe that a slowdown in global warming could be explained by reduced energy receipt at the Earth's surface during a prolonged quiet spell for the Sun. Solar activity at the beginning of the twenty-first century was the lowest for a century (Stauning, 2014) and some predictions suggest that levels in the 2030s could fall to values not seen since the Little Ice Age of the seventeenth century (Shepherd et al., 2014). This could have masked warming due to anthropogenic forcings over the same period. Global cooling by volcanic eruptions

TABLE 1.1 A checklist for meteorological site evaluation. Adapted from WMO (2008)	
Guidelines	Pass
1. The site is not in a downtown area but has municipal services (power supply).	
2. The site is identified in development plans by local government to safeguard against construction of new roads or buildings (to ensure continuity of records).	
3. The site is not near a highway.	
4. The site is expected to remain in stable tenure for at least a century.	
5. The site is representative of its surroundings and has 4000–5000 m^2 area without any depression, hollow, hill, outcrop, etc. nearby.	
6. The site is not subject to any microclimatic influences (induced by topography, coastline, katabatic flows, wind shadowing, etc.).	
7. The site is not vulnerable to natural hazards (such as fluvial flooding, storm surge, lightning strike, wildfire, subsidence or high water table).	
8. The site is not near a building, railway, wall, riverside, large artificial water body or any place that produces heat.	
9. The ground of the observing park is not asphalt, concrete or stone covering. There is natural plant cover such as short grass, weed etc.	
10. The instruments installed in the observing enclosure would not be affected by shadows from buildings, trees, etc.	
11. Any obstacle is at least twice its height (ideally four times its height) distant.	
12. The site commands a wide view of the sky and surrounding country.	
13. The site is on level ground, and will hold an enclosure of at least 25 × 25 m.	
14. The site is accessible to technicians, students and authorities.	
15. There is no sporting activity around the station.	

cont.

TABLE 1.1 (cont.)

Guidelines	Pass
16. The site is not close to any irrigated area (for agriculture, landscaping, etc.).	
17. The site is tidy, well maintained and free from trash.	
18. The site is secure (perhaps near a post office or police station).	
19. The site will be readily accessible to assigned observers.	
20. High-resolution aerial photography, maps and satellite imagery are available for the site area.	

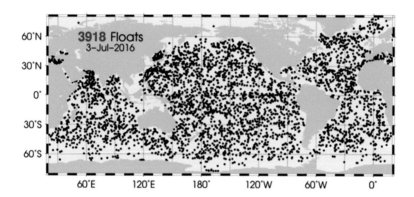

FIGURE 1.5 Number and distribution of Argo floats. *Source*: International Argo Programme http://www.jcommops.org/board?t=Argo [accessed 25/07/16]

(which includes aerosol effects in the lower stratosphere) since 2000 is estimated to be -0.19 ± 0.09 W/m^2, which is equivalent to a reduction in GMST of 0.05–0.12 °C (Ridley et al., 2014). Hence, recent anthropogenic (GHG and sulphur) emissions combined with lower forcing by solar insolation and episodes of volcanism appear to have combined in ways that diminished net radiative forcing.

Ocean Monitoring and Heat Content

Thanks to the Argo programme there are now unprecedented amounts of data for the marine environment. By July 2016, the Argo fleet comprised 3918 drifting floats collecting near real-time data on ocean temperature, salinity and currents (Figure 1.5). Each float

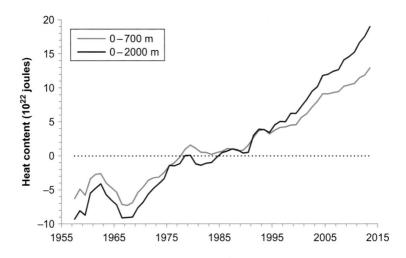

FIGURE 1.6 Global ocean heat content (0–700 and 0–2000 m layers). *Data source*: NOAA

dives to depths of 1000–2000 m before returning to the surface every 10 days to transmit measurements via satellite to Global Data Assembly Centers in Brest, France, and Monterey, California. The Argo array is particularly relevant to the global warming pause because the data are helping to quantify heat storage at different depths in the global ocean. According to the IPCC (2013a:8), "*ocean warming dominates the increase in energy stored in the climate system, accounting for more than 90% of the energy accumulated between 1971 and 2010.*" Although the heat gain at the near surface (0–700 m) may have slowed between 2003 and 2010, heat gain at depth (700–2000 m) has continued unabated since 1993 (Figure 1.6). Overall, it appears that the deep global ocean has buried about one-third of all warming in the 0–2000 m column since 1955 (Leviticus et al., 2012). Has the missing heat been found?

Role of the Pacific and Atlantic Oceans

The scientists' task now is to explain how the seas influence global temperatures, and why the heat is concentrated in certain ocean basins. Kosaka and Xie (2013) believe that the apparent slowdown could be linked to cooling of the tropical Pacific Ocean, which produces cold La Niña-like impacts on global weather. This can occur in the wake of El Niño-rich decades when an immense amount of heat is released from the eastern and central Pacific. Global temperatures rise quickly during periods when the Pacific Decadal Oscillation (PDO) index is positive and the eastern Pacific is warm. Conversely, warming stagnates during spells when the PDO index is negative. This is because stronger Pacific trade winds favour greater subsurface ocean heat uptake through increased subduction in Pacific shallow overturning cells accompanied by equatorial upwelling in the central and eastern Pacific (England et al., 2014). Hence, the pause could reflect natural variations in ocean temperatures driven by the PDO, which switches between warm and cool phases every 15–30 years (Tollefson, 2014).

Attention is also turning to the different mechanisms involved in sequestering heat at great depths in the Atlantic (Kintisch, 2014). Periods with rising GMST (1980–2000) are characterised by heating of the upper Atlantic Ocean; during the hiatus (post 2000) there has been stronger heating of the ocean below 300 m. One theory proposes that as northward moving, tropical saline water encounters the cold atmosphere of the sub-polar Atlantic region it cools, becomes denser, and sinks. In this way, heat is subducted to deeper and cooler but less saline layers of the ocean (Chen and Tung, 2014). Such periods of strong salt-water subduction typically last for 20–35 years and were previously seen in the notably cool 1960s and 1970s. Eventually, the strength of the northward heat transport and salinity anomaly is weakened by greater southward flow of freshwater from ice melt. During these decades surface water in the sub-polar Atlantic becomes cooler, less salty and less dense, so the salinity subduction mechanism slows.

Constructing and Interpreting the Temperature Graph

So, what are the lessons that can be drawn from this brief critique of the GMST record? First, the *prima facie* simplicity of the 'rising temperature graph' belies the amount of data and technical rigour needed to produce it. The salutary lesson is that the accuracy of the most straightforward of climate variables (temperature) can be confounded by many factors (e.g. instrumentation, observer practices, site of station, artificial heat sources). This is even before any weaknesses in the data are interpolated across different terrains, infilled for missing values, or aggregated to a global mean. If this is difficult for temperature, imagine the challenge of accurately measuring global precipitation or ice volume!

Second, having constructed a GMST record, the question then is what exactly is it showing? Changes in surface temperature certainly have relevance to society. But a two-dimensional temperature series does not fully convey the three-dimensional pattern of planetary heating. Therefore, it is hardly surprising that there are so many explanations for the recent behaviour of the GMST record (Table 1.2). As powerful as the GMST graph may be, it is only telling part of the story, so is an inadequate basis for judging whether the climate system is changing in expected ways to radiative forcing (Chapter 4). Global ocean heat content (GOHC) is a much better indicator of overall heat gain within the Earth system (Pielke, 2003).

Implications of Apparent Stop–Start Global Warming

Finally, variations in the rate of global surface heating could have wider ramifications because many anticipated societal impacts are thermally driven. For instance, it is not sensible to extrapolate former (high) rates of glacier melt into river flows for energy, water and food planning. Likewise, temperature-sensitive changes in species distribution, crop yields, sea level rise, and incidence of human disease, wildfire or atmospheric dust may not proceed at a steady pace. Nor does the GMST translate into regionally uniform patterns of warming – Arctic and Mediterranean 'hotspots' may outstrip the global trend (Seneviratne et al., 2016).

Hence, accepting that near-surface temperatures will not rise at a continuous or uniform rate is an important step towards addressing the associated policy and communication challenges. Financing cuts in GHGs may be harder to sell to a sceptical public and media in periods and places with slower change in temperature. But more comprehensive monitoring of the Earth's energy imbalance (using radiation at the top of the atmosphere and ocean heat content) shows that the planet is steadily accumulating excess energy (albeit in places often remote from human experience). Recent deadly heatwaves in Europe and South Asia further highlight the importance of tracking changes in local temperature *extremes* alongside the global mean.

TABLE 1.2 Summary of explanations for the global warming pause

Explanation	Evidence
There is no pause	The period 2001–2010 was the warmest decade on record; 1998, 2005, 2010, 2014 and 2015 all successively broke records; there has been no significant change in trends; surface air temperatures are not a reliable indicator of the Earth's entire heat content.
The data are uncertain	The GMST record is dominated by the behaviour of records in low latitudes; meteorological station density is uneven; data gap filling is problematic over water and sea ice; possible contamination of some data by urban influences.
Weaker radiative forcing	The pause coincides with a period of low net increase in radiative forcing by natural (solar, volcanic) and anthropogenic (GHGs, aerosols) factors.
Natural variability	Temporary cooling phases are associated with cold phase PDO and La Niña conditions; decades of cooling or no warming are expected even with anthropogenic GHG forcing.
Ocean heating	There is no missing heat; it has just been sequestered into deep waters of the Pacific and Atlantic oceans; modes of variability such as the PDO (see above) reflect alternating phases of heat sequestration to the ocean and later release to the atmosphere.
Multiple factors	A slowdown (not pause) is consistent with less solar activity, enhanced heat uptake by oceans (cool phase of the PDO); uneven meteorological station densities and data processing assumptions confound interpretations of uneven rates of surface warming.

1.2 GROUP EXERCISE: SOURCES OF UNCERTAINTY IN TEMPERATURE MEASUREMENT

NOTES

The purpose of this activity is to debate the different types of uncertainty and error that can enter temperature measurement. Discussions may be prompted by a gallery of images, customised to reflect the background of the group. Some exemplars are provided in Figures 1.7 to 1.16. The group should also consider what steps could be taken to counter uncertainties and/or quality assure meteorological records.

The scene could be set with a description of the 'ideal' conditions for a Stevenson screen or automatic weather station. This would give examples of the many factors that must be considered (e.g. site security, maintenance, proximity to high vegetation and buildings, influence of artificial heat sources, instrument position, and representativeness of surroundings). A checklist for meteorological site evaluation is provided in Table 1.1.

FIGURE 1.7 A peri-urban weather station, Loughborough University, UK

FIGURE 1.8 Stevenson screens, Tajikistan. *Photo*: Craig Davies

FIGURE 1.9 Arctic weather station, Storglaciären, Sweden. *Photo*: Tom Matthews

FIGURE 1.10 High-elevation station at Kaskasatjåkka, Sweden. *Photo*: Tom Matthews

FIGURE 1.11 Urban temperature measurement, Athens, Greece.

FIGURE 1.12 Artificial heat sources, John Martin Reservoir, Colorado. *Photo*: Roger Pielke Sr

FIGURE 1.13 Water temperature measured by shielded and unshielded thermistors.

FIGURE 1.14 Meteorological records waiting to be digitised.

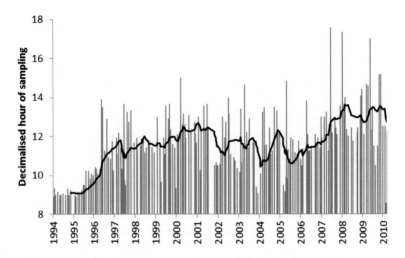

FIGURE 1.15 Dates and times of water temperature measurement at Glutton, River Dove, UK. The black line shows the moving average of 12 samples. Adapted from Toone et al. (2011)

LEARNING OUTCOMES

- To appreciate why the GMST record is so contested, difficult to compile and interpret.
- To recognise the limitations of the GCOS and the range of factors that affect accuracy of temperature measurements at individual sites on land and in the ocean.
- To be able to explain the competing hypotheses for the post 1998 pause in global warming, including denial of existence.

1.3 FURTHER QUESTIONS TO RESEARCH AND DISCUSS

What are the main sources of uncertainty affecting GMST estimates AND how have these changed since the work of Callendar (1938)?

What are the relative strengths and weaknesses of the GMST record and GOHC as indicators of climate change?

How should uneven rates of global warming be explained to the public?

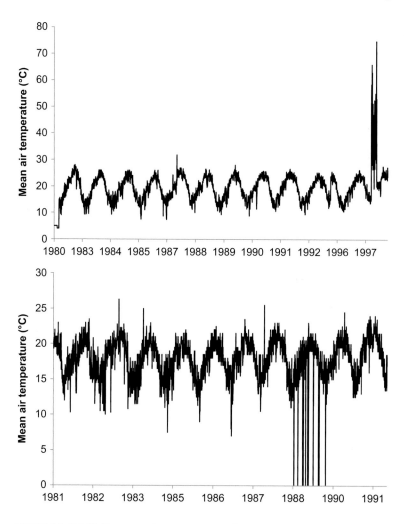

FIGURE 1.16 Daily mean air temperatures for Dumeid (left) and Ibb (right), Yemen.

1.4 FURTHER READING

Kaufmann, R.K., Kauppi, H., Mann, M.L. and Stock, J.H. 2011. Reconciling anthropogenic climate change with observed temperature 1998–2008. *Proceedings of the National Academy of Sciences of the United States of America*, **108**, 11790–11793.

Roemmich, D., Church, J., Gilson, J., et al. 2015. Unabated planetary warming and its ocean structure since 2006. *Nature Climate Change*, **5**, 240–245.

Tomlinson, C.J., Chapman, L., Thornes, J.E. and Baker, C. **2011.** Remote sensing land surface temperature for meteorology and climatology: a review. *Meteorological Applications*, **18**, 296–306.

1.5 OTHER RESOURCES

Argo global array of temperature/salinity profiling floats www.argo.ucsd.edu/About_Argo.html

CRUTEM4 global historical near-surface air temperature anomalies over land www.metoffice.gov.uk/hadobs/crutem4/

CRUTEM4 in Google Earth https://crudata.uea.ac.uk/cru/data/crutem/ge/

Met Office Reports on the global warming pause www.metoffice .gov.uk/research/news/recent-pause-in-warming

NOAA Ocean Heat and Salt Content www.nodc.noaa.gov/OC5/3M_ HEAT_CONTENT/

WMO Global Climate Observing System www.wmo.int/pages/prog/ gcos/

2

Why Does Climate Change?

TOPIC SUMMARY

Weather describes the condition of the atmosphere – as pressure, temperature, precipitation, wind speed, cloud cover, sunshine and humidity – at a point in time. Weather varies over seconds to months whereas climate is the 'average weather' over decades. Climate variability and change are observed across different time scales. Over seasons to years, there are fluctuations in ocean temperatures in the Pacific and Atlantic basins that set the tempo for droughts and widespread flooding across the globe. Decades to centuries of cooling coincide with periods of greater volcanism and lower sunspot activity. Onset and retreat of ice ages over past millennia were driven by subtle variations in the Earth's orbit with changes in receipt of solar radiation at certain latitudes, as well as by large emissions of carbon dioxide (CO_2) from the Southern Ocean.

Since the Industrial Revolution, a growing signature of anthropogenic GHG forcing of the Earth's radiative budget has blended with all these natural drivers. Whereas rising concentrations of CO_2 might once have been regarded as beneficial for averting the next ice age, the modern consensus is that human-induced climate change poses an existential threat. The United Nations Framework Convention on Climate Change (UNFCCC) and Intergovernmental Panel on Climate Change (IPCC) have been central to international efforts to drive down emissions and avoid dangerous climate change.

BACKGROUND READING

O'Hare et al. (2005) provide a synthesis of techniques for reconstructing past climates and explain the main natural and human causes of climate change. Oppenheimer and Petsonk (2005) chart the emergence of the idea of 'dangerous anthropogenic interference' with the climate system and show how it is interpreted by Article 2 of the UNFCCC.

2.1 NATURAL AND HUMAN CAUSES OF CLIMATE
Evidence of Climate Variability and Change

"The only constant about climate is change". Not my words but the sage view of Dr Patterson of Carleton University in Canada. His insight comes from thousands of years of climate, inferred from variations in the abundance of planktonic species preserved in lake deposits. Each organism has a favoured environment – some prefer salinity, others acidic, cool or warm water. Their changing abundance and communities track the ups and downs of lake levels and, with it, climate change. Other evidence of a restless climate is found in tree rings, landforms, and isotopes locked into ice sheets and ocean sediments. Documentary records, ship logs and even artworks yield valuable information about the climate before systematic weather measurements began. Seventeenth-century paintings of the frozen River Thames have, for example, shaped a whole mythology around the Little Ice Age (Jones, 2008). Collectively, such 'proxy' climate records give clues about the scale and causes of variability over decades to millennia. They also help to place the most recent episode of global warming (during the so-called Anthropocene[1]) within a much longer perspective.

Seasonal to Decadal Variability

Climate variability over seasons to decades is generated by two-way energy, mass and momentum transfers between the ocean and atmosphere systems. Over time some of these chaotic exchanges

---

[1] A term accredited to the Nobel scientist Paul Crutzen to describe the epoch during which human activities have made a significant impact on Earth systems.

organise themselves into recurrent patterns of regional air and water circulation such as the El Niño Southern Oscillation (ENSO). When a mass of warm water emerges in the tropical Pacific, predictable areas of drought (northeast Brazil, eastern Australia, Indonesia and India) and flooding (Peru, Ecuador, southwest and southeast United States) generally follow. When the water is cool, most of these weather patterns reverse (e.g. drought in Peru). Such teleconnections between sea surface temperature (SST) anomalies in the Pacific and extreme weather elsewhere are the basis for seasonal forecasting (Chapter 8).

Other, less infamous modes of variability are found in the lexicon of global and regional patterns (Table 2.1). For example, when in positive phase, the North Atlantic Oscillation (NAO)[2] brings relatively mild/wet winters (Figure 2.1a) across large swathes of northwest Europe, the Mediterranean and even parts of Central Asia (Thompson and Wallace, 2001). Similarly, the Pacific Decadal Oscillation (PDO) shows a strong winter temperature dipole over North America with warmth in the northwest coinciding with cold air in the southeast (Figure 2.1b). When positive PDO and ENSO phases coincide an amplified precipitation signature can emerge that is characterised by aridity in the northwest and very wet conditions in the southwest United States. Likewise, the PDO in conjunction with the Atlantic Multidecadal Oscillation (AMO) in SSTs have been linked to below average rainfall over North America, with some 'megadroughts' in the Great Plains lasting decades to centuries (Cook et al., 2009).

Centennial to Millennial Variability

Research by the Past Global Changes project is casting new light on continental-scale temperature variations over the last 2000 years (PAGES 2k Consortium, 2013). A complex picture is emerging of air temperatures that do not vary synchronously across all regions (Figure 2.2). Terms such as the Medieval Warm Period (MWP) and Little Ice Age (LIA) might invoke images of worldwide warmth or cooling but these perceptions have, in the past, been shaped by data and scholarship biased towards the Northern Hemisphere. When worldwide palaeoclimate data are analysed, the exact timing of peak warm and cool phases is found to vary by region and not always in

[2] Also known as the Northern Hemisphere Annular Mode (NAM)

TABLE 2.1 Leading modes of climate variability and their dominant periodicity

Mode	Definition	Periodicity (years)
Antarctic Oscillation (AAO) also known as the Southern Annular Mode (SAM)	Oscillation in the surface atmospheric pressure gradients and associated speed of the upper westerly vortex around the South Pole.	5–7
Atlantic Multi-decadal Oscillation (AMO)	A coherent pattern of variability in SSTs across the North Atlantic basin.	60–80
Arctic Oscillation (AO)	As for the AAO but around the North Pole.	0.5–3
El Niño–Southern Oscillation (ENSO)	Oscillations in the state of the ocean–atmosphere system in the Pacific equatorial region, manifested by warm (El Niño) and cold (La Niña) surface water phases.	2–7
Indian Ocean Dipole (IOD)	Oscillation in the SSTs of the Indian Ocean. During positive (negative) phases SSTs and precipitation in the western Indian Ocean are above (below) average whereas the eastern Indian ocean is cooler (warmer) and drier (wetter) than average.	1.5–10
North Atlantic Oscillation (NAO)	Oscillation in the state of the ocean–atmosphere system in the North Atlantic. During positive	5–8

cont.

Mode	Definition	Periodicity (years)
TABLE 2.1 (cont.)		
	(negative) phases there are large (small) pressure differences between the Azores High and the Atlantic Low near Iceland with strong (weak) westerly winds and advection of the Gulf Stream.	
Pacific Decadal Oscillation (PDO)	Oscillation in SSTs in the North Pacific, manifested by positive (negative) phases when waters are anomalously warm (cold) along the Pacific coast yet cold (warm) in the North Pacific.	20–30

concert with the global temperature trend. Although the period AD 830–1100 was generally warm in the Northern Hemisphere, sustained warmth in Australasia and South America did not emerge until AD 1160–1370. Likewise, a transition to colder conditions began first in the Arctic, Asia and Europe between AD 1200 and AD 1500, but all regions were cold by about AD 1580 except Antarctica, which was relatively warm (Figure 2.2).

Explanations for the late-sixteenth-century cooling include: reduced solar irradiance or output from low sunspot activity; increased volcanism and concentrations of aerosols in the atmosphere; land cover changes such as deforestation; and lower insolation due to variations in Earth's orbit (Crowley, 2000; Bauer et al., 2003; Jones and Mann, 2004). Fluctuations in solar radiation output are linked to the number of sunspots, which varies on average over 11-, 22-, 80–90-year and longer time scales, whereas the length of individual sunspot cycles fluctuates between 10 and 12 years (Figure 2.3). Periods with fewer sunspots and slower cycles have coincided with notable cool interludes, such as the LIA, which was

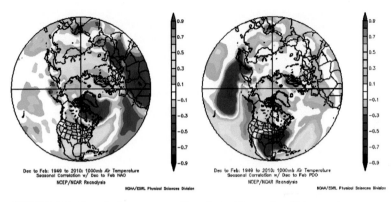

FIGURE 2.1 Correlation patterns for winter (December–February) near-surface air temperatures with the (a) North Atlantic Oscillation (NAO), and (b) Pacific Decadal Oscillation (PDO) indices 1949–2010. Note the positive correlations (yellow to red shading) over northwest Europe (in a) and over Alaska/ northwest Canada (in b). Plots were created with the NOAA climate analysis tool: www.esrl.noaa.gov/psd/data/correlation/

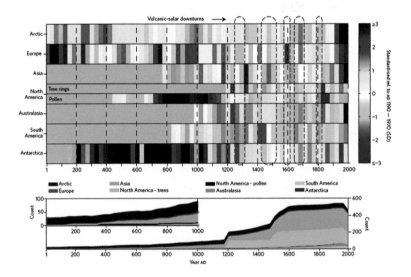

FIGURE 2.2 Continental-scale temperature reconstructions. Thirty-year mean temperatures for the seven PAGES 2k Network regions, standardised to have the same mean and standard deviation over the period of overlap among records (AD 1190–1970). North America includes a shorter tree-ring-based and a longer pollen-based reconstruction. Dashed outlines enclose intervals of pronounced volcanic and solar negative forcing since AD 850. The lower panel shows the running count of number of individual proxy records by region. *Source*: PAGES 2k Consortium (2013)

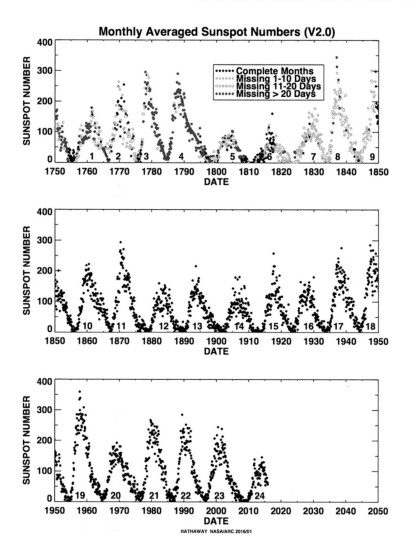

FIGURE 2.3 Monthly average sunspot count and cycle number 1750–2016.
Source: NASA http://solarscience.msfc.nasa.gov/SunspotCycle.shtml
[accessed 05/07/16]

harshest during the well-known Maunder sunspot minimum of 1645 to 1715. A low stage in the sunspot regime is one amongst many explanations offered for the purported global warming pause between 1998 and 2014 (Stauning, 2014) (see Chapter 1).

Volcanic Eruptions and Climate

The amount of solar radiation reaching the Earth's surface also depends on the transparency of the atmosphere. Explosive volcanic eruptions (e.g. Tambora 1815, Krakatoa 1883, Santa Maria 1902, Novarupta 1912, Mount St Helens 1980 and Pinatubo 1991) cool the surface by adding vast quantities of sulphur dioxide to the stratosphere. Atmospheric processes then transform the gas into tiny particles or aerosols that linger and reflect radiation for several years. For example, the Mount Pinatubo eruption injected ~10 million tonnes of sulphur into the stratosphere and caused a global cooling of ~0.5 °C up to 18 months later (Soden et al., 2002). Climate model simulations of major eruptions show strengthened stratospheric winds and Atlantic overturning circulation, as well as lingering signatures in ocean temperatures (Raible et al., 2016). After the immense Krakatoa eruption, a cool layer of water was subducted into the deep ocean, where it is still detectable as a thermal anomaly (Gleckler et al., 2006).

Orbital Variations and Climate

Climate reconstructions for the Last Interglacial (LIG) period around 125,000 years ago suggest that maximum temperatures were 2–5 °C higher than present. Ocean sediments further indicate typical rates of sea level rise of 1.2 m/century during this time (Berger, 2013), compared with ~0.3 m/century presently. Variations in climate over these geological time scales correlate with changes in the latitudinal and seasonal distribution of solar insolation caused by Milankovitch cycles in the stretch, tilt and wobble of the Earth's orbit (Figure 2.4). Resulting changes to surface energy receipt, in turn, trigger major feedbacks in the climate system involving ocean temperatures, sea ice extent, vegetation cover and land ice. For example, increased summer insolation caused by orbital forcing between 22 and 18 thousand years ago is thought to have reduced the area of sea ice and hence albedo of the Southern Hemisphere. Less reflected solar radiation served to warm the ocean further and ushered in the onset of deglaciation (WAIS Divide Project Members, 2013). Warming was further assisted by CO_2 venting from the depths of the Southern Ocean during a period with weaker North Atlantic overturning circulation (Meckler et al., 2013).

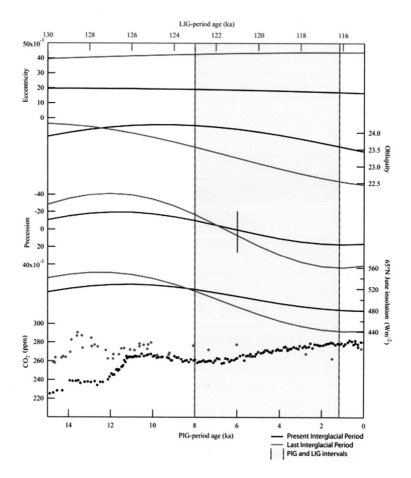

FIGURE 2.4 Climate forcings during the Present Interglacial Period (PIG, black lines) and Last Interglacial Period (LIG, red lines). Incoming top of the atmosphere solar radiation is affected by the eccentricity (stretch) of the Earth's elliptical orbit, obliquity (tilt) of the Earth's axis, and precession of the equinoxes (wobble) of the North Pole orientation. Oxygen isotope signatures held in long sediment cores show the rhythm of these Milankovitch cycles to be 23,000 years for precession, 41,000 years for obliquity, and 96,000 years for eccentricity. *Source:* Bakker et al. (2014)

Anthropogenic Greenhouse Gas Emissions and Climate

The current (post-1974) warm phase is unusual because it is the only period in the last 1000 years during which both hemispheres have synchronous positive temperature anomalies (Neukkom et al., 2014).

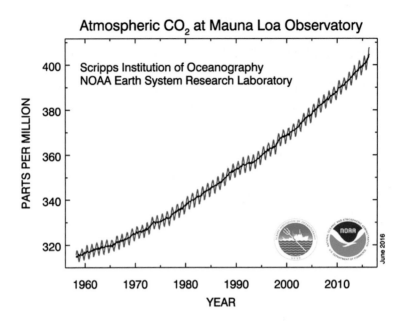

FIGURE 2.5 Monthly mean atmospheric CO_2 concentration at Mauna Loa, Hawaii. *Source*: Dr Pieter Tans, NOAA/ESRL (www.esrl.noaa.gov/gmd/ccgg/ trends/) and Dr Ralph Keeling, Scripps Institution of Oceanography (scrippsco2.ucsd.edu/) [accessed 05/07/16]

This is attributed to human activities that are changing regional and global climate by altering the composition of the atmosphere and land surface properties. By the 1860s, Tyndall had established that CO_2 (and other GHGs such as nitrous oxide, methane and water vapour) selectively absorbs solar radiation in the infrared band. At that time the atmospheric concentration of CO_2 was about 280 ppm; by 2014 the monthly mean concentration had surpassed 400 ppm (Figure 2.5). Year-on-year fossil fuel combustion and land use changes are strengthening heat absorption by adding GHGs to the atmosphere at an accelerating rate. In 2014 annual global emissions of CO_2 were ~35 Gt compared with ~21 Gt in the early 1990s (see Figure 12.1).

More than a century ago Arrhenius calculated that doubling the concentration of CO_2 would change the global mean surface temperature (GMST) by 4–5 °C (Crawford, 1997). The Fifth Assessment Report of the Intergovernmental Panel on Climate Change (IPCC,

2013a:16) stated: "*equilibrium climate sensitivity*[3] *is likely in the range 1.5°C to 4.5°C (high confidence), extremely unlikely less than 1°C (high confidence) and very unlikely greater than 6°C (medium confidence).*" The amount of warming is ambiguous because of large uncertainties about future aerosols and ocean heat uptake, as well as complex feedbacks in the climate system (e.g. from changing cloud, snow or ice cover). Some scientists now assert that the upper end GMST changes presented in the IPCC (2013a) report are too high and inconsistent with observed warming (Stott et al., 2013). Others claim that the transient climate response to CO_2 may be underestimated because climate models sequester heat more efficiently than the real ocean (Kuhlbrodt and Gregory, 2012).

Other Human Drivers of Climate

There is growing appreciation that human modifications of land cover also contribute to climate change. For example, forest clearance, crop cultivation and soil degradation release carbon to the atmosphere. Between 1990 and 2005, estimated net emissions of carbon[4] from land use change were 1500 ± 700 MtC/yr (compared with 8700 ± 700 MtC/yr from fossil fuel combustion in 2008) (Le Quéré et al., 2009). Conversion of natural vegetation into pasture or arable land contributes methane emissions from cattle and submerged fields, or nitrous oxide from fertilisers. Climate feedbacks from forest die-back or thawing tundra areas could further add to the burden of GHG emissions. Another concern is that human destruction or degradation of terrestrial and marine environments affects their capacity to act as carbon sinks. Sabine et al. (2004) estimate that about one-third of the long-term carbon storage potential of the ocean has already been taken by historical anthropogenic emissions.

Although the global land area classified as urban is only 3% (Liu et al., 2014), it is well known that cities exert a strong influence on local temperatures compared with surrounding rural landscapes. This is because built areas retain more solar heating during the

---

[3] This is the expected global warming caused by doubling the effective CO_2 concentration. 'Effective' is used because the sensitivity implicitly captures forcing by other GHGs. The value of the sensitivity parameter is important because it strongly determines the amount of climate change projected by climate models and is highly contested (see Chapters 3 and 4).

[4] To convert C to equivalent CO_2 emissions multiply the former by 3.67.

day, and have lower rates of radiant cooling during the night. Urban areas also have lower wind speeds, less convective heat loss and evapotranspiration, which leaves more solar energy for local surface heating. Air conditioning, space heating, transport, cooking and industrial activity add to this thermal load. With rapid growth of urban populations, particularly in developing countries, megacity regions are fast becoming hotspots of GHG emissions (see Chapter 10). In 2009 the C40 cities[5] accounted for 291 million people and contributed 1747 Mt of CO_2, ranking the group fourth most populous and carbon emitting if treated as a nation (Hoornweg et al., 2011). The high population densities and concentration of assets are also increasing human vulnerability to climate hazards (Wilby, 2007). Hence, cities are now regarded as major players in both climate mitigation (Chapter 12) and adaptation (Chapter 14) efforts.

Changing Climates of Fear

The realisation that human activities could adversely change the global climate is less than 50 years old. Until the 1970s, the prevailing fear was of a return to ice age conditions. Pioneers in the field such as Arrhenius (1908) famously claimed: *"By the influence of the increasing percentage of carbonic acid [carbon dioxide] in the atmosphere, we may hope to enjoy ages with more equable and better climates."* Even Callendar (1938:236), who was the first to correlate changes in GMST with CO_2, thought that rising atmospheric concentrations of the gas would delay the *"return of the deadly glaciers"*. Now, opponents to action on GHG emissions emphasise the societal benefits of elevated CO_2, such as enhanced biological productivity leading to larger crop yields (e.g. Goklany, 2015).

However, the overwhelming scientific consensus is that human activity has contributed to the observed GMST rise since pre-industrial times and that future climate change presents an existential threat to human and natural systems. The Third Assessment Report of the IPCC (2001) identified five major reasons for concern (see Chapter 11). The United Nations Framework Convention on Climate Change (UNFCCC) was the first international treaty intended

---

[5] A group of megacities committed to tackling climate change; see www.c40cities.org/

to avert 'dangerous' climate change by setting a path towards binding reductions in GHG emissions (see Esmaeili, 2010). Although the accomplishments of the Kyoto Protocol and subsequent accords have been modest in terms of cutting emissions, the journey has highlighted that ameliorating the human causes of climate change is as much a moral–ethical matter as a political–economic or scientific–technical challenge.

Major Themes of Climate Change

The rest of this book explores the practicalities of addressing climate change under four main themes. (1) The **causes** – scientific understanding and mathematical representation of the climate system at global to local scales, moving into ways in which models are used to support decision-making (Chapters 3 to 6). (2) The **consequences** – evaluation of climate change impacts with special attention to national risk assessment, the water sector, urban and rural areas, and developing regions (Chapters 7 to 10). (3) The **countermeasures** – steps that can be taken to decarbonise economies, to adapt to threats and opportunities, or even geoengineer Earth to avoid 'dangerous' climate change (Chapters 11 to 15). (4) The **communications** – ways in which climate change is framed and conversed by experts, sceptics and climate service providers (Chapters 16 to 17). Finally, climate change is considered alongside other global challenges in Chapter 18.

2.2 GROUP EXERCISE: CLIMATE CHANGE QUIZ

NOTES

This activity introduces some important concepts about climate science and policy that are unpacked later in the book. The quiz is intended to raise awareness in a light-hearted way. Some questions may appear quirky but these are meant to challenge assumptions or make a point. The choice of case study countries and statistics (here mainly UK) should be modified as required. The quiz works best when participants form small teams. It is suggested that the list of questions is divided into 'rounds', giving groups an opportunity to compete.

Round 1: Warm up

1. Which of the following is not a greenhouse gas?	(a) Water vapour (b) Oxygen (c) Carbon dioxide (d) Methane
2. What percentage of all carbon dioxide emissions since 1750 has remained in the atmosphere?	(a) 0.04% (b) 0.4% (c) 4% (d) 40%
3. What was the mean atmospheric concentration of CO_2 in 1750 and in 2014?	(a) 280 and 310 ppm (b) 280 and 398 ppm (c) 280 and 575 ppm (d) 180 and 283 ppm
4. Which of the following gases has the longest residence time in the lower atmosphere?	(a) Carbon dioxide (b) Water vapour (c) Methane (d) Sulphur hexafluoride (SF_6)
5. Radiant heat from an adult is how many times stronger than all human warming and cooling influences on the climate (net ~1.6 W/m^2)?	(a) 40× (b) 10× (c) 2× (d) The same

Round 2: Past, present and future

6. The Earth has warmed by approximately how much since 1850?	(a) 4.1 °C (b) 2.3 °C (c) 1.0 °C (d) 0 °C
7. How many of the top 10 warmest years since 1850 have occurred after 1998?	(a) 9 (b) 7 (c) 5 (d) 3
8. Why is the climate outlook for coming decades so uncertain?	(a) Future greenhouse gas concentrations are unknowable

(b) Future solar output and
volcanic aerosols are unknown
(c) Science and models of the
Earth system are imperfect
(d) All of the above

9. To avoid 'dangerous' climate change the European Union says we must limit global warming to what?	(a) 0.5 °C (b) 1 °C (c) 2 °C (d) 5 °C
10. Which country in 2013 had the worst combination of low readiness and high vulnerability to climate change	(a) UK (b) Ethiopia (c) Chad (d) Norway

Round 3: Footprints

11. On average, in 2012 how much CO_2 did each UK citizen create?	(a) 1.6 tonnes (b) 7.1 tonnes (c) 7.7 tonnes (d) 16.4 tonnes
12. Which of the following activities generates the most CO_2 emissions?	(a) 12,000 miles driving (b) Family of four flies roundtrip London to Los Angeles (c) One year of average meat consumption (d) Drinking a can of cola each day
13. How much more efficient is a low-energy light-bulb (compared with a conventional bulb)?	(a) 80% (b) 50% (c) 20% (d) Less than 5%
14. What percentage of national CO_2 emissions is due to the treatment and supply of all the water used in England and Wales?	(a) 10% (b) 5% (c) 2% (d) 1%

15. The energy saved from recycling one glass bottle would power a TV for how long?
    (a) 2 mins
    (b) 20 mins
    (c) 2 hours
    (d) 20 hours

Round 4: Low carbs

16. Switching to a renewable electricity supplier saves the average UK household how much CO_2 per year?
    (a) 100 kg
    (b) 500 kg
    (c) 1000 kg
    (d) 2000 kg

17. Electrical gadgets on stand-by in the UK produce as much CO_2 as how many transatlantic air passenger journeys?
    (a) 14,000
    (b) 140,000
    (c) 1,400,000
    (d) 14,000,000

18. Which of the following contributes most to the UK's carbon footprint?
    (a) Recreation and leisure
    (b) Space heating
    (c) Clothing
    (d) Commuting

19. What is the UK Government target for CO_2 emission reductions by 2050 (relative to 1990)?
    (a) 100%
    (b) 80%
    (c) 40%
    (d) 20%

20. The equivalent of how many new nuclear power stations are needed for the UK to be on track for its 2050 emission reduction target?
    (a) 30
    (b) 10
    (c) 3
    (d) 1

LEARNING OUTCOMES

- To be able to describe the dominant natural drivers of climate variability at seasonal, annual, decadal, centennial and millennial time scales.

cont.

- To recognise that reconstructed natural temperature variations exhibit asynchronous change between and within hemispheres, whereas the era of human-induced climate change shows more synchronous planetary warming.
- To appreciate that rising atmospheric concentrations of CO_2 have not always been perceived as 'dangerous' to human and natural systems.

2.3 FURTHER QUESTIONS TO RESEARCH AND DISCUSS

What has been the most important scientific contribution to the understanding of anthropogenic climate change? Justify your choice.

What are the dominant natural and human drivers of climate change over decadal time scales?

Produce a table showing the key data sources for describing climate variability at decadal, centennial and millennial scales.

2.4 FURTHER READING

Hulme, M. 2009. *Why We Disagree About Climate Change: Understanding Controversy, Inaction and Opportunity*. Cambridge University Press, Cambridge.

Fleming, J.R. 1998. *Historical Perspectives on Climate Change*. Oxford University Press, Oxford.

2.5 OTHER RESOURCES

Intergovernmental Panel on Climate Change (IPCC) www.ipcc.ch/

NOAA Earth System Research Laboratory Climate Indices www.esrl.noaa.gov/psd/data/climateindices/list/

NOAA Mauna Loa observatory CO_2 data http://co2now.org/Current-CO2/CO2-Now/noaa-mauna-loa-co2-data.html

NOAA Solar Indices Data www.ngdc.noaa.gov/nndc/struts/results?t=102827&s=1&d=8,4,9

United Nations Framework Convention on Climate Change (UNFCCC) http://unfccc.int/2860.php

3

What Does It Take to Build a Model of the Climate System?

TOPIC SUMMARY

Climate models are mathematical representations of key Earth system components (e.g. ocean, atmosphere, biosphere, cryosphere and hydrosphere). They are used to simulate the complex behaviour of the system under natural and human-induced changes to the energy balance. Climate models were not always this sophisticated, having evolved in line with available supercomputing technologies, Earth observations and scientific theory. Simpler models can still provide useful insights about system feedbacks and thresholds (tipping points) beyond which there are more dramatic climate responses to radiative forcing. The case for further climate model development is now predicated on economic cost–benefit analysis, which emphasises utility of model predictions to society. However, internal model climate variability means that there is considerable uncertainty about the future outlook for regional temperatures and precipitation.

BACKGROUND READING

McGuffie and Henderson-Sellers (2014) and Edwards (2011) provide synopses of climate model development. Shukla et al. (2010) set out an ambitious manifesto for climate modelling that transcends national research and computing capabilities. Marotzke and Forster (2015) use a comparatively simple statistical model to explore the relative importance of radiative forcing, climate feedback, ocean heat uptake and internal variability in global mean surface temperature trends.

3.1 RECIPE FOR A CLIMATE MODEL
Do It Yourself Climate Modelling

Scientists like to talk about their research. So imagine the delight whenever a political leader asks a climate expert to explain his/her work. This was the case in early 1990 when Warren Washington, one of the founding fathers of climate modelling, had a private meeting to discuss global warming with Governor Sununu (President George H.W. Bush's White House Chief of Staff). By Washington's (2008) own account, much of the conversation centred on the role played by the ocean in controlling rates of climate change. Apparently the Governor was sceptical about climate models and wanted to be able to perform his own analyses. To this end, Washington left the meeting with a request from Sununu to build a climate model that would run on an office computer.

The task presented a significant technical challenge because, at that time, the available Compaq 386 had just 0.2% of the computing power of an iPhone 6! Nonetheless, Washington and colleagues at the National Center for Atmospheric Research (NCAR) set about reducing tens of thousands of lines of computer code from their three-dimensional global climate model into a simpler one-dimensional, atmosphere–ocean version. The scientists recognised that the job presented a unique opportunity to inform an influential climate policy-maker. It took the NCAR team a few weeks to build and test the model, and to add a graphical interface. The whole scheme was designed to complete a simulation in minutes. The climate model was duly delivered to the White House on a single floppy disk, and that was the last that the group heard of it.

This encounter illustrates the factors that are still important to climate model development today: available computing power, plausible representations of physical processes, and relevance to society. Each vital ingredient is elaborated below.

Computing Power

According to Moore's Law, computer speeds are expected to double nearly every 18 months. Climate model evolution is intrinsically linked with these advances and would now be impossible without supercomputing facilities (McGuffie and Henderson-Sellers, 2001). For example, in 2004 the UK Met Office NEC SX-6 supercomputer was capable of performing 1.9 trillion calculations per second to

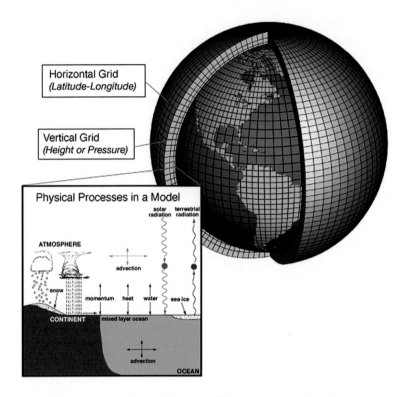

FIGURE 3.1 Schematic of a climate model. *Source*: NOAA http://
celebrating200years.noaa.gov/breakthroughs/climate_model/modeling_
schematic.html [accessed 05/07/16]

represent the global climate system via a grid of cells each covering
40 km × 40 km. Move forward a decade to 2014 and their Cray XC40
supercomputer was expected to perform 16,000 trillion calculations
per second with local grid cells as small as 1.5 km × 1.5 km. This is
the coarsest spatial resolution at which summer convective storms
can be represented over the UK.

Extra computing power can be deployed in a variety of ways
depending on scientific and operational priorities. Doubling the
horizontal and vertical resolution of a climate model increases the
computational burden by about a factor of ten (Figure 3.1). Another
option is to use the available power to conduct more stringent
analysis of model uncertainties by performing a greater number
of simulations with slightly different initial conditions of the
ocean–atmosphere state, or with a wider range of values for uncertain parameters in the model. A third possibility is to run model

simulations over centuries rather than decades to assess the long-term reversibility of climate change (e.g. Solomon et al., 2009). In practice, climate modelling centres allocate their resources across various experiments that trade off spatial resolution with model complexity, or with number or length of runs according to research goals.

Representing Physical Processes in Code

The immense computational burden of climate models is due to the growing fidelity and complexity of simulations. For example, the atmosphere component of the UK Met Office HadGEM2-CC model has a horizontal grid resolution of ~250 km and 60 vertical layers to represent an altitude of 84 km (Collins et al., 2011; Martin et al., 2011). For this part of the climate model alone, differential equations for mass, energy, moisture and momentum conservation must be solved for ~1.66 million grid cells at each model time step (normally 10–20 minutes) for scenarios out to year 2100. The fragment of computer code shown in Table 3.1 gives some insight to the complexity of these algorithms. The sample is part of a scheme to represent convection in the atmosphere. An entire climate model assembled from atmosphere, ocean, land, ice, carbon cycle and dynamic vegetation components would consist of millions of lines of Fortran 90 computer code like this. Such is the complexity that no solitary climate scientist has command of the entire model and, with decades of development, one family of models could have been built by hundreds, even thousands of researchers from very diverse disciplines.

First Generation Climate Models

The earliest ocean–atmosphere climate models were developed in the late 1960s. Manabe and Wetherald (1975) performed the first doubling of atmospheric CO_2 experiment with a general circulation model (GCM) with 500 km resolution and nine vertical levels to represent the structure of the planetary boundary layer and stratosphere of the Northern Hemisphere (Figure 3.2). Ocean had to be treated as a 'swamp' with infinite moisture supply but no heat transport via marine currents. Solar and long-wave radiation fluxes were estimated from observed patterns of annual mean cloudiness. Distributions of water vapour and ozone were determined by

TABLE 3.1 A fragment of Fortran code for the Miller convection scheme within a climate model. *Source*: Mike Blackburn www.met.reading.ac.uk/~mike/betts_miller/ndeep4v2 [accessed 13/07/16]

```
C
C                      SECOND ITERATION FOR MOIST ADIABAT.
C
        ZQSATC=C2ES*EXP(C3LES*(ZTC(JL,JK)-TMELT)*
     *                 (1./(ZTC(JL,JK)-C4LES)))/APP1(JL,JK)
        ZCOR=1./MAX(ZEPCOR,(1.-VTMPC1*ZQSATC))
        ZQSATC=ZQSATC*ZCOR
        ZQCD=(ZQC(JL,JK)-ZQSATC)/(1.+C5LES*ZCONS2*ZQSATC*ZCOR*
     *                 (1./(ZTC(JL,JK)-C4LES))**2)
        ZTC(JL,JK)=ZTC(JL,JK)+ZCONS2*ZQCD
        ZQC(JL,JK)=ZQC(JL,JK)-ZQCD
      ENDIF
C
C                      PARCEL TEMPERATURE FOR BUOYANCY.
C                      *ZGAMMA* IS A MIXING FRACTION FROM CLOUD-TOP
C                      ENTRAINMENT.
C
      ZZQ=MAX(ZEPQ,ZQP1(JL,JK))
      ZP1=ZZQ*APP1(JL,JK)/(ZC6+ZC7*ZZQ)
      ZTSP=ZC2+ZC3/(ZC4*ALOG(ZTP1(JL,JK))-ALOG(ZP1)-ZC5)
      ZZSP=APP1(JL,JK)*XLG((ZTSP/ZTP1(JL,JK)),ZC1)
      ZRATDP=(APP1(JL,NLEV)-ZZSP)/(APP1(JL,NLEV)-APP1(JL,JK))
      ZAPP1=APP1(JL,IBASE(JL))
      ZTCLD=ZTC(JL,JK)*(1.0-ZGAMMA*ZRATDP)+ZGAMMA*(ZTP1(JL,JK)+
     *      ZTCB(JL)*XLG((APP1(JL,JK)/ZAPP1),ZCONS1)*
     *      (ZRATDP-1.0))
```

equations that linked radiation and hydrological cycle; the CO_2 concentration was assumed to be the same everywhere. Their water vapour calculations also accounted for evaporation, vertical mixing, convection and condensation; soil moisture and snow cover were derived from water balance estimates for continental surfaces. The soil heat balance depended on albedo (the reflectivity of the surface) which, in turn, was related to latitude and snow–ice cover (permanent and temporary).

Available supercomputing resources in the 1970s meant that this GCM could only be run twice for an 800-day simulation of

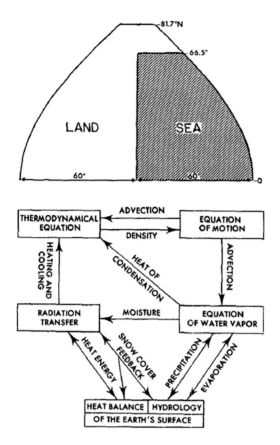

FIGURE 3.2 The distribution of continent and 'ocean' (top) and major components (bottom) of a first generation general circulation model. *Source:* Manabe and Wetherald (1975) ©American Meteorological Society. Used with permission.

contemporary CO_2 concentrations and once for doubled CO_2. The last 100 days of each run were used to estimate the quasi-equilibrium state of the atmosphere. The global mean change in temperature associated with doubling CO_2 was found to be 2.93 °C. Despite the simplicity of the GCM (relative to the complexity of modern Earth system models, Figure 3.3) the double-CO_2 warming still falls comfortably within the range of the IPCC Fifth Assessment Report (Figure 3.4). Therein lies a paradox that climate modellers have to confront: why do successive generations of climate models, with increasing levels of complexity, arrive at roughly the same global mean temperature response to given

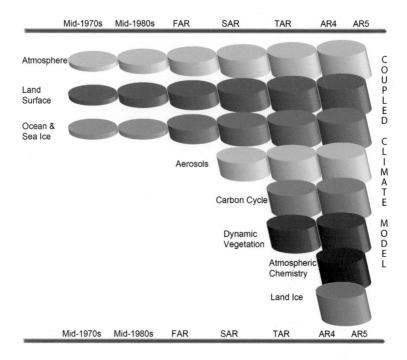

FIGURE 3.3 The development of climate models over the last 35 years showing how the different components were coupled into comprehensive climate models over time. In each aspect (e.g. the atmosphere, which comprises a wide range of atmospheric processes) the complexity and range of processes has increased over time (illustrated by growing cylinders). Note that during the same time the horizontal and vertical resolution has increased considerably, e.g. for spectral models from T21L9 (roughly 500 km horizontal resolution and 9 vertical levels) in the 1970s to T95L95 (roughly 100 km horizontal resolution and 95 vertical levels) at present, and that now ensembles with at least three independent experiments can be considered as standard. *Source*: Cubasch et al. (2013)

GHG forcing? This leads to a second question: what value has been added by investments in climate modelling over the last 40 years?

Climate Model Evolution and Use

The spread in global mean temperature change or spatial patterns of temperature and precipitation variability barely changed between the climate model ensembles used in IPCC AR4 (CMIP3) and AR5 (CMIP5) despite substantial model development. As Knutti and

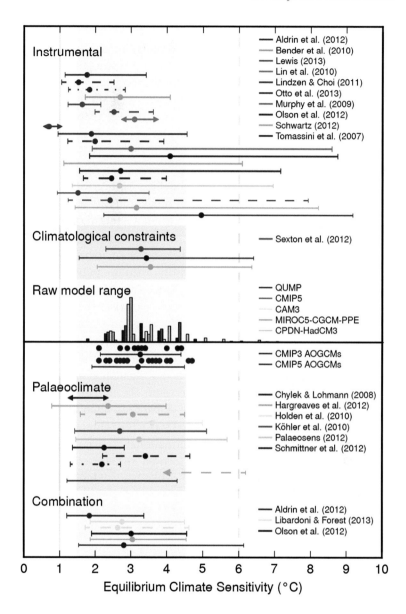

FIGURE 3.4 Distributions and ranges for equilibrium climate sensitivity based on different sources of evidence (e.g. observations, climate models, palaeoclimate data). The grey shaded range marks the likely 1.5–4.5 °C range, grey solid line the extremely unlikely less than 1 °C, the grey dashed line the very unlikely greater than 6 °C. *Source*: IPCC (2013b)

Sedláček (2012) observe, much of the increased supercomputing power was invested in building more complete models rather than increasing the resolution of existing models. Hence, through the relatively recent incorporation of carbon-cycle feedbacks and atmospheric chemistry, the present climate modelling community is more confident that most significant processes are now included. They further assert that much of the model spread (Figure 3.4) is explained by the irreducible uncertainty of the internal variability of the climate system (see Chapter 2).

As the scale and cost of climate modelling have grown, so have pressures to demonstrate the utility of model outputs for decision-making and society. Gone are the days when a handful of scientists could build a model in a few weeks that would run on a politician's office computer. [Although the *ClimatePrediction.net* experiment has ushered in an era of distributed computing with the public running simplified versions of the Met Office climate model on home computers.] A hallmark of the climate research papers of the 1970s was their quest for *process insight* and representation (e.g. Schneider, 1972). Now there has been a subtle shift in emphasis. Opening statements of journal articles are more likely to stress the *significance* of this process understanding to society (e.g. Kendon et al., 2014).

At some point in the late 1970s the remit for climate modelling also expanded from understanding the global consequences of rising GHG emissions to assessing the regional impacts of climate on socio-environmental systems. Calls for multinational research and computing facilities are now very much focused on delivering climate services and "*policy-relevant climate predictions*" (Shukla et al., 2010:1412). For example, the business case for a new Met Office supercomputer[1] claimed that improved weather forecasting would save 74 lives and £260 million per year, whilst the anticipated benefits of improved preparation and contingency planning for climate change could be worth £2 billion to the UK economy.[2]

Utility for Adaptation Planning

Others question whether climate models are really needed for adaptation planning (Dessai et al., 2005) or even fit for this purpose

[1] www.publications.parliament.uk/pa/cm201012/cmselect/cmsctech/1538/153802.htm
[2] www.metoffice.gov.uk/news/releases/archive/2014/new-hpc

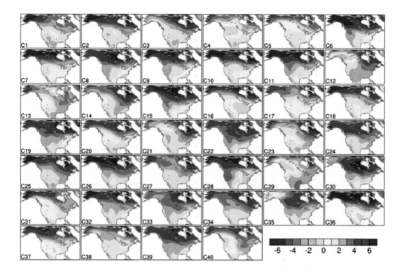

FIGURE 3.5 Winter surface air temperature trends for the period 2010–2060 (expressed as °C per 51 years) based on a single climate model (CCSM3) given 40 different initial states of the atmosphere. *Source*: Deser et al. (2014) ©American Meteorological Society. Used with permission.

(Kundzewicz and Stakhiv, 2010). This is partly because regional climate model projections are very sensitive to the conditions used to start experiments. For example, one study ran a single climate model 40 times, beginning each experiment with minutely different atmospheric states but with the same GHG forcing over the period 2010–2060 (Deser et al., 2014). The results show that very large uncertainty in winter air temperature trends over North America can arise just from natural climate variability within the model (Figure 3.5). Whilst some runs (C6, C15, C26, C36) predicted strong warming (>5 °C) over Canada, others (C4, C16, C28, C36 and C40) actually show cooling over parts of the United States despite increased GHG concentrations. Large initial conditions uncertainty was confirmed by the same authors using another climate model. Spatial variations in seasonal precipitation trends were found to be even more marked than temperature trends.

Cycles of Climate Model Development

So what is the recipe for a twenty-first century climate model experiment? Take a big team of dedicated research scientists with expertise in chemistry, climatology, computer science, geography, hydrology, mathematics, oceanography, physics and statistics. Over the course of several decades, incrementally add important physical climate components such as the carbon cycle or dynamic ice sheets (Figure 3.3). Invest approximately £100 million in 140 tonnes of supercomputing hardware and feed it 2.5 MW of electricity (equivalent to the energy consumption of more than 5000 UK households). Take the best available observations of the latest ocean temperatures and atmospheric conditions to start the experiment. Use the half million computer processors to perform 16 quadrillion calculations per second to produce climate change simulations to the end of the twenty-first century at cloud-resolving scales. Throughout the six months that the experiment takes to run, record snap-shots of the climate state to the available 17 billion gigabytes of storage. Spend up to a decade evaluating the model compared with observed atmosphere, terrestrial and ocean data. Share the results with other climate model centres and likewise cross-check against their models. Translate the stored climate model outputs into formats that can be used by other research groups and consultants whilst communicating key messages to stakeholders and policy-makers. After five years, present the business case for the next model and supercomputer upgrade. . .

3.2 GROUP EXERCISE: UNPICKING *DAISYWORLD*

NOTES

This activity uses basic energy balance equations to demonstrate some fundamental processes and behaviours of climate models. Recall from Chapter 2 that positive feedbacks amplify responses (e.g. atmospheric temperature–water vapour content, ice sheet albedo–temperature interaction, tundra warming–methane release) and that negative feedbacks dampen reactions (e.g. cloud albedo–temperature relations) to internal and external radiative forcings. Feedbacks can lead to abrupt, non-linear changes in climate as well as emergent properties such as self-regulation. This has led some to conclude that *since the climate system is complex, occasionally chaotic, dominated by abrupt changes and driven by competing feedbacks with largely unknown thresholds, climate prediction is difficult, if not impracticable* (Rial et al., 2004:30).

Daisyworld (Lovelock and Margulis, 1974) is used here to explore the dynamic responses of a simple climate system to different

cont.

initial conditions, parameter settings and external forcing via an Excel version of the model (adapted from Hardisty et al., 1993). *Daisyworld* also provides a vehicle for discussions about theories linking system complexity to system stability, as well as about how a basic climate model might be enhanced.

Daisyworld was devised by James Lovelock not to predict the future but to illustrate a key concept at the heart of his Gaia theory: interactions between the Earth's living (biotic) and non-living (abiotic) components create a self-regulating system (or "*atmospheric homeostasis by and for the biosphere*") (Lovelock and Margulis, 1974). Their imaginary planet is inhabited only by black and white daisies. The growth rate of each is controlled by local temperature (Figure 3.6); conversely each species affects local temperature because the amount of solar radiation absorbed depends on daisy colour (surface albedo). The reflected portion of incoming sunlight is only 25% for black daisies, 75% for white daisies and 50% for bare ground. Given the initial area covered by each species and their death rate, it is possible to simulate the changing fraction of the planet covered by the different daisies as well as emergent albedo–temperature responses to increasing solar luminosity.

Using default settings (Table 3.2) the area occupied by black daisies peaks before that of the white daisies (Figure 3.7, top). However, the amount of bare ground and temperature remain stable across a wide range of incoming solar radiation (forced by increasing solar luminosity). Initial local warming and more rapid growth of black daisies are gradually replaced by stable temperatures and expansion of white daisies under more intense solar radiation (Figure 3.7, bottom). Nonetheless, a point is reached beyond which even the white daisies are unable to regulate the energy receipt so temperatures rise abruptly. At this threshold luminosity white daisies become extinct and the planet has 100% bare ground. Overall, the biotic control system of the living planet stabilises temperatures compared with steady warming on a lifeless planet.

The basic model has since been extended to include grey daisies, herbivore (rabbit) and predator (fox) dynamics, genetic mutation, and biological amplification of weathering (a geochemical negative feedback) (Lenton, 1998).

Worksheet

Download an Excel version of *Daisyworld* from here:
www.cambridge.org/wilby

Spend a few minutes familiarising yourself with the initial model parameters. Values shaded yellow in the spreadsheet can be adjusted:

TABLE 3.2 *Daisyworld* model parameters		
Parameter	Value	Notes
Proportion of planet suitable for growth	1.00	Maximum area that can be colonised
Initial area covered by white daisies	0.20	Proportion at first time step
Initial area covered by black daisies	0.20	Proportion at first time step
Albedo of white daisies	0.75	Proportion of solar radiation reflected
Albedo of black daisies	0.25	Proportion of solar radiation reflected
Albedo of bare ground	0.50	Proportion of solar radiation reflected
Death rate	0.30	Proportion of daisies lost each time step
Solar constant	1000.00	Energy from the Sun
Luminosity	0.70	Proportion of present day value (1.0)
Stefan's constant	0.000000057	Relates black body radiation to temperature
Constant to relate temperatures and albedo	20.00	Temperature change per unit albedo change

PART A

Refer to the example plots or adjust the model parameters to determine:

1. What is the temperature response to increasing the luminosity of the Sun over the range 0.7–1.7 on a lifeless and living planet?

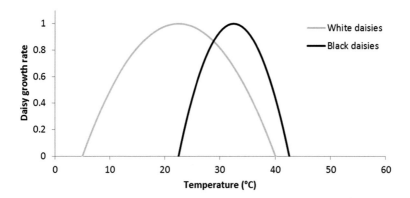

FIGURE 3.6 White and black daisy temperature–growth rate curves. Here, white daisies grow between 5 °C and 40 °C but most rapidly at 22.5 °C. Black daisies have a narrower tolerance and grow at temperatures between 22.5 °C and 42.5 °C, but most rapidly at 32.5 °C.

2. What is the temperature response of a planet inhabited by a single species (i.e. set the initial area of black or white daisies to zero)?
3. What is the effect of adjusting the albedo of black and white daisies?
4. How might Earth's land albedo change under hotter/drier conditions?

PART B

Connect to the online *Daisyball* programme here: http://gingerbooth .com/flash/daisyball/

5. What is the effect of varying the number of daisy species between 1 and 16?
6. What insights to Earth system modelling are gained from *Daisyworld* and *Daisyball*?
7. Suggest three changes to *Daisyworld* that would improve model realism.

LEARNING OUTCOMES

- To appreciate that climate model development is closely tied to technical advances in supercomputing power, Earth observation and evolving scientific understanding of physical processes.
- To know that despite vastly increased complexity and realism of physical processes contained in climate models, the uncertainty in projected global mean temperature under

cont.

double CO_2 concentrations is still in the region 1.5–4.5 °C, just as it was for much simpler models in the 1970s.

• To appreciate that climate modelling is a costly, multi-disciplinary endeavour that has to be justified in economic terms by the range of climate services and benefits afforded to society.

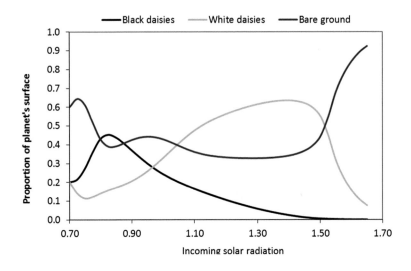

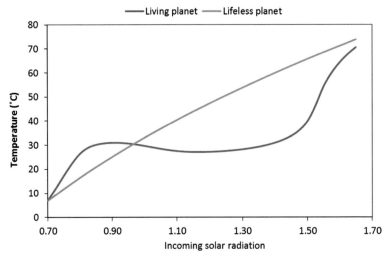

FIGURE 3.7 Proportion of the planet surface that is bare ground or covered by daisies (top) and a comparison of the temperature response for a planet with (living) and without daisies (lifeless) under increasing levels of incoming solar radiation (bottom).

3.3 FURTHER QUESTIONS TO RESEARCH AND DISCUSS

What is the most essential ingredient of climate modelling? Justify your answer.

What are the main obstacles to global climate modelling in developing countries?

What might be the advantages and disadvantages of a multinational climate research and computing facility analogous to the CERN (*Conseil Européen pour la Recherche Nucléaire*) particle physics laboratory in Geneva?

3.4 FURTHER READING

Hawkins, E. and Sutton, R. 2010. The potential to narrow uncertainty in projections of regional precipitation change. *Climate Dynamics*, **37**, 407–418.

Shapiro, M., Shukla, J., Brunet, G., et al. 2010. An Earth-system prediction initiative for the twenty-first century. *Bulletin of the American Meteorological Society*, **91**, 1377–1388.

Stainforth, D.A., Aina, T., Christensen, C., et al. 2005. Uncertainty in predictions of the climate response to rising levels of greenhouse gases. *Nature*, **433**, 403–406.

3.5 OTHER RESOURCES

ClimatePrediction.net www.climateprediction.net/ [accessed 05/07/16]

Climate Modelling 101 http://nas-sites.org/climate-change/climatemodeling/ [accessed 05/07/16]

Climate Models (WMO) www.wmo.int/pages/themes/climate/climate_models.php [accessed 05/07/16]

Monash University Simple Climate Model http://monash.edu/research/simple-climate-model/mscm/index.html [accessed 05/07/16]

Wilby, R.L. 2011. Imagine a world without climate models. http://pielkeclimatesci.wordpress.com/2011/01/17/imagine-a-world-without-climate-models-by-rob-wilby-of-loughborough-university/ [accessed 05/07/16]

4

How Trustworthy Are Climate Models?

TOPIC SUMMARY

Climate model realism is judged by the models' ability to represent the mean, trend, variability and extremes of the observed
climate. More elaborate consistency tests determine whether
models can conserve mass or simulate phenomena such as El
Niño. These checks are not straightforward to perform because
there is often as much uncertainty in observed climate data as in
the modelled climate. Even when models agree with observations
and with each other there is no guarantee that their predictions
will be accurate. This is because uncertainty surrounds the sensitivity of the climate to future aerosol and GHG forcing. For
instance, partially understood Earth system feedbacks involving
methane release or the carbon cycle could increase future climate
sensitivity compared with the present.

BACKGROUND READING

Covey et al. (2003) demonstrate a range of techniques and indices
for comparing climate models with observations. The paper also
includes an example of a Taylor (2001) diagram that simultaneously compares the errors, variability and pattern skill of models
against observations. Reichler and Kim (2008) present a skill index
that combines 14 variables to assess objectively the ability of three
generations of model to simulate the present mean climate.

4.1 BEAUTY IS IN THE EYE OF THE BEHOLDER
Models Inform Costly Decisions

According to the eminent statistician George Box, "*all models are wrong, but some are useful*". This is fine if models are confined to the laboratory. But when they shape global policies and investments, or even life and death decisions, weighing the accuracy of model predictions becomes a very serious matter. For instance, the ability to mitigate climate change depends on a host of factors (not least political willingness and development of low-carbon energy sources) but costs could be in the region of a 0.06 percentage point reduction in growth rates (IPCC, 2014). Put another way, *Energy Technology Perspectives 2014* estimate that US$44 trillion of extra investment is needed to decarbonise the global economy by 2050 to limit climate change to 2 °C of warming. (Some claim that the potential fuel savings and avoided climate damages would more than offset this cost.[1])

Climate Sensitivity to Greenhouse Gas Forcing

Climate models predict rates of global warming and attendant impacts given atmospheric concentrations of GHGs and land surface properties. As a result, model parameters such as climate sensitivity to doubled CO_2 concentrations assume critical importance in economic evaluations of alternative decarbonisation pathways. For instance, high-end climate sensitivity (4.5 °C under doubled CO_2) implies rapid warming and the need for early, more aggressive abatement of fossil fuel emissions to avoid 'dangerous' climate change (see Chapter 13). Conversely, low-end climate sensitivity (1.5 °C under doubled CO_2) with emissions near current levels would postpone crossing the dangerous 2 °C global warming threshold by about a decade (Rogelj et al., 2014). With such profound ramifications for climate policy, no wonder that the value assigned to climate sensitivity has become a battle ground for those proposing and opposing urgent action on mitigation (e.g. Lewis and Crok, 2014).

Comparing Models with Observed Climate

If one parameter (climate sensitivity) is problematic, imagine the colossal task of checking the trustworthiness of climate models as

[1] www.iea.org/Textbase/npsum/ETP2014SUM.pdf

a whole. Recall that the latest generation of models are built from vast computer programs that simulate the behaviour of Earth system components at millions of points representing different levels of ocean and atmosphere (Figure 3.1). However, even the observations against which the models are to be checked are some-times of questionable quality (Chapter 1). Given the highly loaded policy implications of climate model projections there is clearly an urgent need for stringent evaluation criteria. Wilby (2010a) suggests five guiding principles:

1. Judge model skill against observation skill – differences between models and observations may actually be smaller than differences between different sets of observations (i.e. observational data are uncertain too).
2. Ensure that the comparison is like for like – model output is in gridded format so compare with gridded observations at the same temporal and spatial scale.
3. Apply measures of model skill that are meaningful to decision-makers – users of climate model information will then better understand the value (or not) of what they are receiving.
4. Recognise that climate model skill is just one link in a long chain of uncertainty stretching from the economic drivers of emissions to regional climate impacts – uncertainty due to natural climate variability or impact estimation may swamp errors from climate models in the short term.
5. Test model skill using near-term (seasonal to decadal) forecasts – alternatives such as waiting until the 2050s to confirm a model prediction are not helpful.

Comparing Time Series

Conventional approaches to climate model evaluation rely on time series and pattern correlation techniques (see below). In the first case, the shapes of graphs from different models and observations are expected to overlap, at least from one decade to the next. At first glance, comparison of observed and CMIP5 simulated global mean surface air temperature *anomalies* appears to corroborate the models (Figure 4.1a). However, closer inspection of the right-hand axis of the plot reveals that *actual* simulated temperatures can depart from the observed mean by nearly 1.5 °C, or even more in simpler Earth system models (Figure 4.1b). This might not appear

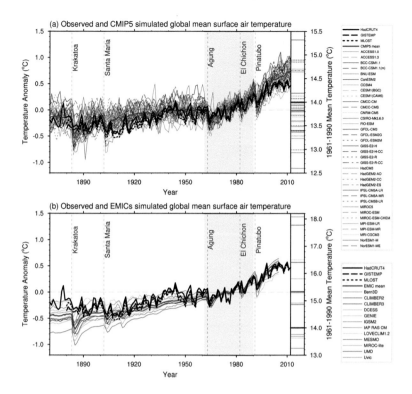

FIGURE 4.1 Observed and simulated time series of the anomalies in annual global mean surface temperature. All anomalies are differences from the 1961–1990 time mean of each individual time series. The reference period 1961–1990 is indicated by shading; vertical dashed grey lines represent times of major volcanic eruptions. (a) Single simulations for CMIP5 models (thin lines); multi-model mean (thick red line); different observations (thick black lines); (b) As in (a) but for Earth system models of intermediate complexity (EMICs). *Source*: Flato et al. (2013), Figure 9.8

a cause for concern until the physical consequences of such biases are considered. First, the bias shown is for the global average, so local errors could be much larger (Figure 4.2b). Second, for sites with mean temperatures close to 0 °C, a warm bias could convert a snow-covered surface to a snow-free one; the same warm bias might be sufficient to change the vegetation from tundra to steppe. Both of these temperature-dependent responses could trigger land–atmosphere feedbacks in the model by modifying the local albedo (i.e. surface reflectivity), radiant energy receipt and water balance.

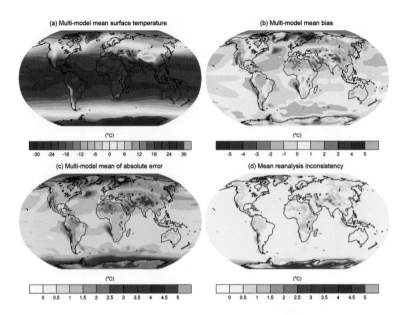

FIGURE 4.2 Annual-mean surface (2 m) air temperature (°C) for the period 1980–2005. (a) Multi-model (ensemble) mean constructed with one realisation of all available models used in the CMIP5 historical experiment. (b) Multi-model-mean bias as the difference between the CMIP5 multi-model mean and the climatology from ECMWF reanalysis of the global atmosphere and surface conditions (ERA)-Interim. (c) Mean absolute model error with respect to the climatology from ERA-Interim. (d) Mean inconsistency between ERA-Interim, ERA 40-year reanalysis (ERA40) and Japanese 25-year reanalysis (JRA-25) products as the mean of the absolute pairwise differences between those fields for their common period (1979–2001). *Source*: Flato et al. (2013), Figure 9.2

Comparing Spatial Patterns of Climate

Pattern correlation methods involve grid-to-grid comparisons of observed and modelled quantities (Figure 4.2a). The expectation is that a good model will place the hot and cold (or wet and dry) places in the same areas as the observed climate. Models do reproduce the global distribution of annual mean surface temperature with very high confidence. In this case, the pattern correlation is $r\sim0.99$ despite biases of several degrees in some high-altitude regions (e.g. Himalayas and Andes), along ice margins in the North Atlantic, and in zones of ocean with cool water upwelling close to the equator. Overall, there is a tendency for a cold bias in the models but notable 'hotspots' are evident in central Asia, over the Amazon basin and in North America above the Arctic circle (Figure 4.2b). Note also that absolute biases in the CMIP5 models are larger than absolute differences between two versions of observed temperature (compare Figure 4.2c with 4.2d). The pattern correlation is weaker for precipitation (r~0.82; Figure 4.3) but has improved since the IPCC Fourth Assessment Report (Flato et al., 2013). Large errors are known to occur at regional scales, most notably over the Himalayas, Rockies, Sahel, equatorial Pacific and Atlantic (Figure 4.3d).

Present Skill Does Not Equal Future Skill

Care is needed when interpreting biases estimated from large numbers of models because errors could be cancelling out within the ensemble (Figures 4.2b and 4.3b). Knutti (2008) cautions that

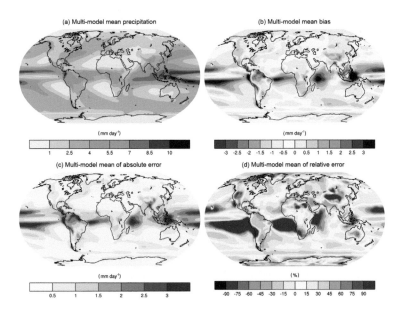

FIGURE 4.3 Annual-mean precipitation rate (mm day^{-1}) for the period 1980–2005. (a) Multi-model mean constructed with one realisation of all available AOGCMs used in the CMIP5 historical experiment. (b) Difference between multi-model mean and precipitation analyses from the Global Precipitation Climatology Project (Adler et al., 2003). (c) Multi-model-mean absolute error with respect to observations. (d) Multi-model-mean error relative to the multi-model-mean precipitation itself. *Source*: Flato et al. (2013), Figure 9.4

climate models can be tuned to fit twentieth-century global mean temperatures by adjusting interdependent parameters governing radiative forcing by aerosols and climate sensitivity. When aerosol cooling is assumed to be high, then climate sensitivity to CO_2 has to be high to compensate, and vice versa. Hence, different combinations of these parameters can yield similar behaviour in the calibration period, but different scenarios when applied under higher GHG concentrations in the future. In other words, climate models may simulate observed patterns of warming but for very different reasons (Crook and Forster, 2011).

Future Feedbacks

Major feedbacks and non-linear responses in the climate system also mean that skill at simulating present climate does not guarantee the accuracy of future climate projections (Rial et al., 2004). Whereas there is high confidence that the water vapour feedback amplifies changes in climate, there is more uncertainty about the net radiative feedback due to clouds (especially low clouds). Carbon cycle feedbacks are expected to augment anthropogenic CO_2 in the atmosphere by contributing emissions from soils and vegetation changes (e.g. carbon released from Amazon forest die-back). Likewise, thawing tundra and wetlands could add more methane. What remains less clear is the *strength* of these feedbacks: too weak and the rate of modelled global warming would be slower than observed; too strong and the models would exaggerate the pace of change. Feedbacks may also amplify regional climate change. For example, Seneviratne et al. (2013) show that parts of the Mediterranean region that are projected to have decreasing precipitation and drier soils also have decreased latent heating and increased sensible heating. This soil moisture–climate feedback could augment local maximum daytime temperatures by up to 2.5 °C and reduce summer precipitation by 30–50% (which favours yet more drying and heating).

Faith in Physics

Randall et al. (2007:601) assert that *"confidence in climate models comes from their physical basis, and their skill in representing observed climate and past climate changes"*. Others claim that important terms in the

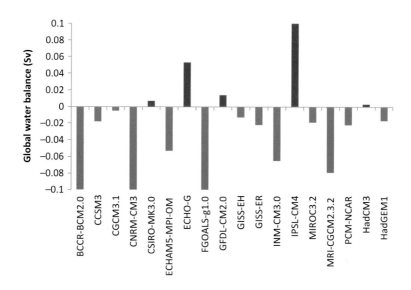

FIGURE 4.4 Global annual mean errors in the atmospheric water balance of CMIP3 climate models. One sverdrup (Sv) of water is $10^6\,\mathrm{m^3/s}$ or $31{,}600\,\mathrm{km^3/yr}$. Note that four climate models have residuals of >0.1 Sv which is approximately 10% of the observed atmospheric moisture transport from ocean to land at any point in time (1.2 Sv). Blue bars denote models that compute more precipitation than evaporation; red bars show models with more evaporation than precipitation. *Data source*: Liepert and Previdi (2012)

Earth's energy budget can be constrained using measurements of surface temperature, ocean heat uptake and satellite observations of radiative fluxes (e.g. Murphy et al., 2007). Nonetheless, consistency tests reveal gross biases when climate models are evaluated against the laws of physics. For example, Liepert and Previdi (2012) showed that the majority of climate models used in CMIP3 generate more global annual mean precipitation than evaporation – what they termed 'ghost' precipitation (the blue bars in Figure 4.4). In these experiments the law of mass conservation is broken, raising fundamental questions about the credibility of the models (if there is more rain than evaporation that cannot be correct). Even if this error is neglected, there are still consequences for the rest of the climate system. The extra water in the soil affects the allocation of heat for surface warming and evaporation, and introduces an artificial, non-radiative forcing of $\sim$0.2 W/m^2. When the same water cycle consistency test was applied to CMIP5 still only 18 out of 30 models were physically consistent (Liepert and Lo, 2013). Climate models are also known to produce precipitation more often and more lightly than is observed (Stephens et al., 2010), or to exaggerate the influence of soil moisture feedbacks on convective rainfall (Taylor et al., 2012).

Multi-faceted Consistency Tests

The El Niño Southern Oscillation (ENSO) is the dominant mode of global climate variability on seasonal to inter-annual scales (Table 2.1). In order to simulate this complex phenomenon, climate models have to be able to represent component processes such as deep convection, wind strength, short-wave radiation and cloud feedbacks over the tropical Pacific. Unfortunately, this is not yet the case and, as a consequence, CMIP5 models show errors in observed ENSO amplitude, period and spatial patterns of precipitation (Bellenger et al., 2013; Flato et al., 2013).

Other tests reveal that CMIP3 models do not capture the pattern or strength of teleconnections between ENSO and the South Asian (Annamalai et al., 2007) or West African (Paeth et al., 2008) monsoons. Other modes of variability in the Indian and Atlantic oceans have been similarly assessed. For instance, Weller and Cai (2013) report that CMIP5 models generate larger than observed variance in sea surface temperatures and amplitude of the Indian Ocean Dipole (Table 2.1), which casts doubt on their projections of austral spring rainfall in this region. Masato et al. (2013) find that the same ensemble underestimates the frequency of winter and summer atmospheric blocking (i.e. high pressure) over the Northern Hemisphere. These model biases suggest possible underplay of drought risk in the future.

Agreement between Models

Consensus amongst climate models about global or regional patterns of change should not be mistaken for confidence in their projections. This presupposes that each model is an independent representation of the climate system and thereby contributing additional useful information (Pirtle et al., 2010). In practice, climate models share structural features such as the same fundamental equations for conservation of momentum, parameters, sub-grid schemes, or modules representing the land surface. Therefore, a degree of similarity is expected a priori, not least because of the common physics included in models. International collections of climate models are, therefore, unlikely to be truly independent, so the effective number of models in the sample is actually smaller than might appear at face value. Hence, there is a risk of double counting and undue confidence whenever models produce similar patterns of climate change (Knutti, 2010; Masson and Knutti, 2011).

Accounting for Natural Climate Variability

Climate model evaluation is further complicated by the intrinsic randomness of the climate system and incomplete representation of known forcings. For example, other important climate drivers beyond carbon dioxide (e.g. aerosol effects on clouds, black carbon deposition, reactive nitrogen, and dynamic land cover) are just beginning to be included in Earth system models (Figure 3.3). Multi-decadal climate model experiments that omit these known factors would not be expected to replicate observed regional climate changes (Pielke et al., 2009). Discussions surrounding the global warming pause have focused greater attention on the ability of climate models to simulate changes in ocean heat content (see Chapter 1). Closer scrutiny of this aspect of the models has revealed that most have less stratified oceans than is observed (Kuhlbrodt and Gregory, 2012). This means that heat is transported more efficiently and buried in deeper layers of the ocean than in reality. With this higher mixing efficiency the rate of global warming in the model would be slower but the contribution of ocean thermal expansion to sea level rise would be enhanced.

Summary of Findings

Climate model evaluation is a technically demanding and important activity. Many issues can confound model assessment *even for the present climate*. Discrepancies could be due to factors such as spatial and temporal variations in the quality of observations; natural multi-decadal variability; climate model parameter uncertainty; missing or imperfect representation of known forcings. Table 4.1 summarises evidence of climate model skill compiled by the IPCC Fifth Assessment Report. Experts judged the models according to global means, trends, variability and extremes. Model performance was assessed as high for most temperature indices; medium for

TABLE 4.1 Summary of how well CMIP5 models simulate important features of the climate of the twentieth century. Model performance and confidence in the assessment were rated based on expert judgement. Adapted from Figure 9.44 in Flato et al. (2013)

Climate features	Model performance	Confidence in assessment
Mean state		
Sea surface temperatures	High	Very high
Arctic sea ice extent	High	Very high
Surface air temperature	High	High
Global monsoon	Medium	High
Blocking weather patterns	Medium	Medium
Large-scale precipitation	Medium	Medium
Snow albedo feedback	Low	Very high
Global land carbon sink	Low	Medium
Soil moisture	Low	Low
Trends		
Surface air temperature	High	Very high
Global ocean heat content	Medium	High
Antarctic sea ice extent	Low	Medium
Variability		
North Atlantic Oscillation	High	Medium
El Niño Southern Oscillation	Medium	High
Inter-annual CO_2 concentration	Low	Very high
Extremes		
Annual number of Atlantic hurricanes	High	Medium
Droughts	Medium	Medium
Tropical cyclone tracks and intensity	Low	Medium

precipitation and drought patterns; low for terrestrial carbon and tropical cyclones. Confidence in the assessment was generally lower for variability and extremes than for means and trends. This reflects the quality of the observational data and the amount of research undertaken to date. Overall, there was lowest confidence in both modelling and assessment of soil moisture patterns. Following chapters explore the extent to which useful insights can still be gained from climate models despite these recognised shortcomings.

4.2 GROUP EXERCISE: CLIMATE MODEL CONTEST

NOTES

This activity reveals the complicated process of climate model evaluation. Even with a standard set of metrics (some of which are more intelligible than others) it is difficult to establish whether one model is more skilful than another. Parts of the exercise refer to the CMIP5 'mosaic' diagram (Figure 4.5).

Set the scene by sharing some recent news stories about climate models. These will demonstrate that, despite their immense complexity, climate models have already entered mainstream public debates about climate change. To help contextualise the exercise, it would also be helpful to give a few every-day examples of how we routinely evaluate technologies (e.g. via technical specifications for televisions, smart-phones, cars, even sports equipment). Invariably some sifting and trade-offs are involved in the selection of the preferred article.

Then provide an explanation of the relative error scale shown in Figure 4.5: the darker the blue (red) the better (weaker) the model performance. Detailed definition of the variables is less important but it is helpful to point out that different levels in the atmosphere are represented. These range from top of atmosphere (TOA), high-altitude (200 hPa), middle (500 hPa) to lower (850 hPa) atmosphere. Perhaps the two most meaningful quantities are total precipitation (PR) and temperature (TA). The group should debate the questions in small sets then feed back their findings via a facilitated discussion.

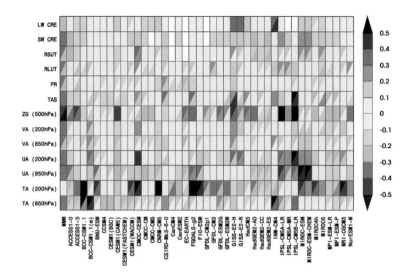

FIGURE 4.5 Relative error measures of CMIP5 model performance, based on the global seasonal-cycle climatology (1980–2005) computed from historical experiments. Rows and columns represent individual variables and models, respectively. The error measure is a space–time root-mean-square error (RMSE), which, treating each variable separately, is portrayed as a relative error by normalising the result by the median error of all model results. For example, a value of 0.20 indicates that a model's RMSE is 20% larger than the median CMIP5 error for that variable, whereas a value of –0.20 means the error is 20% smaller than the median error. White indicates that model results are currently unavailable. A diagonal split of a grid square shows the relative error with respect to both the default reference dataset (upper left triangle) and the alternate (lower right triangle). Relative errors are calculated independently for the default and alternate datasets. *Source*: Flato et al. (2013), Figure 9.7

Key: LW CRE: Clear sky top of atmosphere (TOA) long-wave cloud radiation [W/m^2]; **SW CRE:** Clear sky TOA short-wave cloud radiation [W/m^2]; **RSUT:** TOA reflected short-wave radiation [W/m^2]; **RLUT:** TOA reflected long-wave radiation [W/m^2]; **PR:** Total precipitation [mm/d^1]; **TAS:** Surface (2 m) air temperature [°C]; **ZG:** Geopotential height (at the specified pressure level) [m]; **VA:** Meridional (north–south) wind speed (at the specified pressure level) [m/s]; **UA:** Zonal (west–east) wind speed (at the specified pressure level) [m/s]; **TA:** Temperature (at the specified pressure level) [°C]; **MMM:** Multi-model mean.

Worksheet for Climate Model Contest

Questions for group discussion:

1. What minimum levels of skill might be expected from a climate model?
2. Based on the information provided in Figure 4.5, which are the most (and least) skilfully produced variables? Which are the 'best' (and 'worst') climate models?
3. List some *non-atmospheric* quantities that could be used to test climate models.
4. What *policy-relevant* measures of model performance might be applied?
5. What other factors might affect users' confidence in climate model information?

LEARNING OUTCOMES

- To recognise that judgements about whether a model is trustworthy or not depend on the performance measure and observations used for comparison.
- To appreciate the difference between a 'reality test' involving specific, like-for-like model versus observation and a 'consistency test' for overall compliance with the basic laws of physics (such as conservation of mass).
- To understand that neither consensus between models nor skilful simulation of the present climate are reliable indicators of the accuracy of climate projections.
- To be aware that even small biases in model behaviour can trigger large, non-linear regional climate feedbacks (such as higher temperatures changing the snowline and with it the reflectivity and energy balance of a surface).

4.3 FURTHER QUESTIONS TO RESEARCH AND DISCUSS

Explain the difference between climate model reality and consistency tests.

Why is skill at simulating the present climate a necessary but insufficient basis for accurately projecting the future climate?

What artificial climate feedbacks could occur in models where the precipitation bias is (i) too wet or (ii) too dry compared with reality?

4.4 FURTHER READING

Cowtan, K., Hausfather, Z., Hawkins, E., et al. 2015. Robust comparison of climate models with observations using blended land air and ocean sea surface temperatures. *Geophysical Research Letters*, 42, 6526–6534.

Lutz, A.F., Immerzeel, W.W., Gobiet, A., Pellicciotti, F. and Bierkens, M.F.P. 2013. Comparison of climate change signals in CMIP3 and CMIP5 multi-model ensembles and implications for Central Asian glaciers. *Hydrology and Earth System Sciences*, 17, 3661–3677.

Orlowsky, B. and Seneviratne, S.I. 2013. Elusive drought: uncertainty in observed trends and short- and long-term CMIP5 projections. *Hydrology and Earth System Sciences*, 17, 1765–1781.

4.5 OTHER RESOURCES

Comparing CMIP5 and observations www.climate-lab-book.ac.uk/comparing-cmip5-observations/ [accessed 06/07/16]

Coupled Model Intercomparison Project (CMIP) http://cmip-pcmdi.llnl.gov/

KNMI Climate Explorer http://climexp.knmi.nl/plot_atlas_form.py

Taylor diagrams (tools for comparing climate models with observations) https://climatedataguide.ucar.edu/climate-data-tools-and-analysis/taylor-diagrams

5

What Is the Purpose of Regional Climate Downscaling?

TOPIC SUMMARY

Regional climate downscaling is used to bridge the gap between the coarse resolution of climate model output (~200 km) and the assumed information needs of decision-makers at river basin and city scales (~10 km). Many techniques exist for performing this task, but it is unclear whether scaling relationships observed now will hold under changed climate conditions. Much research has evaluated downscaling methods under the present climate such that their comparative advantages are well known. What remains less clear is whether these techniques actually add value to decision-making processes. Hence, there are growing calls to apply downscaling tools in smarter ways to hazard forecasting, infilling and repair of meteorological data, as well as to the appraisal of adaptation options.

BACKGROUND READING

Wilby and Wigley (1997) provide a critique and early comparison of various downscaling methods. Tang and Dessai (2012) assess the extent to which the UK Climate Projections 2009 (UKCP09) provided downscaled information that was usable for adaptation planning.

5.1 ADDING VALUE TO CLIMATE RISK ASSESSMENT AND PROCESS UNDERSTANDING

Stormy Times Ahead

Imagine that you are an urban drainage engineer appointed to upgrade parts of the nineteenth-century sewerage network of Newcastle upon Tyne. You are aware that heavy downpours in summer 2012 overwhelmed the sewers causing widespread flooding and disruption across the city. You also know that any new or upgraded infrastructure could be in place to the end of the twenty-first century and possibly beyond. You recall that the UK Met Office produced guidance on future daily rainfall return periods for major cities in England and Wales (Sanderson, 2010). Their models suggest that a heavy rainfall event that presently occurs just once in 30 summers could have a return period of anywhere between 7 and 33 years by the 2080s (Figure 5.1). In other words, the design storm[1] could happen more or less frequently! You ask your boss whether, to be on the safe side, you should design a drainage network capable of absorbing a summer rain storm that currently has a return period of 1 in 100 years. Your reasoning is that a downpour of this magnitude could become the 1 in 30 years event by the 2080s (see Figure 5.1, top right panel).

Open Questions about Regional Climate Changes

This is a plausible project. Engineers and planners recognise that the climate does not stand still (Milly et al., 2008) and that regional climate change safety margins need to be factored into long-lived infrastructure design, at least in the UK (Environment Agency, 2011). But, is it reasonable to expect a 25 km resolution climate model to accurately predict extreme summer rainfall likelihoods for the 2080s at the city scale? Does the climate model even simulate realistic convective storms? Can the model reliably capture climate change signals if future land surface modifications and aerosols are ignored? How might the results change if different GHG emissions scenarios or climate models are used? Was the cost and time invested in producing the scenarios really worth it?

---

[1] The largest rainfall total (mm) in a 30-year period that is fully conveyed by the drainage system

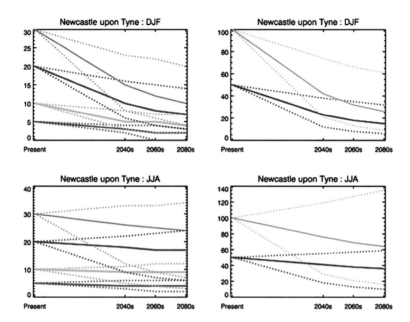

FIGURE 5.1 Change in return period for rainfall events with present-day return periods of 1 in 5 (red), 1 in 10 (green), 1 in 20 (blue), 1 in 30 (orange) (left-hand panels) and 1 in 50 (purple) and 1 in 100 years (grey) (right-hand panels). Return periods (years) are shown on the vertical axis. The central estimate (50th percentile) is indicated by a solid line, and the 10th and 90th percentiles, calculated using the full range of probabilistic projections from UKCP09, illustrate the possible range of return periods and are shown by dotted lines. The present-day return periods are positioned at 1980 on the horizontal axis (marked as 'Present'). Changes for winter (DJF, top row) and summer (JJA, bottom row) have been calculated separately. Note that the scale of the vertical axis is different for each panel. *Source*: Sanderson (2010)

These are some of the important questions being raised by both users and suppliers of high-resolution climate change scenarios.

Rationale for Downscaling

So called 'downscaling' methods are a broad family of techniques for generating fine-scale (typically less than 10–25 km resolution) climate change information from coarse-scale (typically ~200 km resolution) global climate model (GCM) output (Wilby et al., 2009). These procedures correct biases from atmospheric processes or surface features that are too small to be resolved (e.g. convective clouds, small water bodies, coastline) or are neglected at the scale of GCMs (e.g. direct radiative forcing by aerosols) (Gustafson et al., 2011). There are two main types of downscaling: statistical and dynamical (regional climate models, RCMs). Both assume that the empirical relationships or RCM parameters relating the GCM to local scale are unchanging through time (the so-called 'stationarity' assumption). Both are dependent entirely on the accuracy of the information supplied by the host GCM to drive the downscaled behaviour.

Statistical Downscaling Methods

Statistical downscaling (SD) relates large-scale atmospheric features (such as pressure patterns or humidity) to the local variable(s) of interest (such as daily rainfall totals, temperature or sunshine hours) (Figure 5.2). The empirical predictor-to-predictand scaling can be achieved in a variety of ways (Table 5.1). The simplest approach is to remove biases in GCM output by, for example, scaling climate model rainfall amounts to match observations (e.g. Pierce et al., 2015). Another common method is to use a regression model (or other transfer function). Weather typing (or classification) methods assume that the local variables of interest (e.g. surface ozone concentration) depend on concurrent

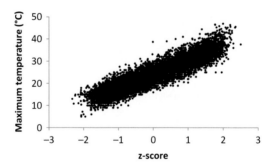

FIGURE 5.2 The relationship between GCM-scale near-surface temperatures (expressed as a z-score) and daily maximum temperatures (°C) in Tunis for the period 1961–2000. With this relationship it is possible to reconstruct past data (as in Figure 5.5). Or the function can be used to project local temperatures given future GCM temperatures, assuming that the historical relationship remains the same. Local temperature influences such as sea breezes or urban expansion are neglected.

TABLE 5.1 Strengths and weaknesses of the four main downscaling methods

Method	Strengths	Weaknesses
Regional climate model (RCM) Three-dimensional models that simulate local climate features given time-varying information about the surrounding climate state from a GCM. The RCM domain is nested within the grid of the GCM. The RCM is typically run off-line, hence any regional climate change does not feedback to the host GCM.	• Derives 1.5–25 km resolution climate information given GCM-scale input • Responds in physically consistent ways to different external forcings • Resolves atmospheric processes such as orographic precipitation • Consistency with host GCM • Yields maps of climate change within the domain	• Depends on the realism of GCM boundary forcing, choice of domain size and location • Requires significant time and computing resources • Sensitive to initial boundary conditions (e.g. soil moisture content) • Non-trivial to transfer to a new region or domain
Weather typing (SD) Models that group local meteorological variables using an objective or subjective classification of the prevailing weather pattern into 'types'. Example classification schemes include the Lamb Weather Types (UK) or the Grosswetterlagen (Europe).	• Yields physically interpretable linkages between weather patterns and surface climate • Versatile (e.g. applicable to surface climate, air quality, flooding, erosion) • Amenable to the analysis of extreme events (e.g. floods and droughts) • Useful for multi-site applications	• Requires additional effort of weather classification • Weather pattern schemes can be insensitive to climate change if atmospheric moisture is neglected • There may be a wide variety of meteorological conditions within a given weather type

cont.

TABLE 5.1 (cont.)		
Method	**Strengths**	**Weaknesses**
Transfer function (SD) Statistical models of observed cause–effect relationships between large-scale variables such as atmospheric pressure and local variables such as rainfall. The form of the relationship may be linear or non-linear, single or multi-site.	• Relatively straightforward to apply • Can employ a wide range of predictor variables to capture local climate response to radiative forcing and dynamical changes • 'Off-the-shelf' tools and software are available	• Poor representation of observed variance and extreme events • May assume linearity and/or normality of data • Results are sensitive to the choice of predictor variables and domain with respect to the target site(s)
Weather generator (SD) Models whose random behaviour does not exactly duplicate weather sequences but resembles the overall statistical distribution of daily weather at a location. Most use rainfall occurrence to infer secondary variables such as temperature, solar radiation and humidity.	• Production of large ensembles for uncertainty analysis or long simulations for extremes • Model parameters can be interpolated for data-sparse regions using site properties such as elevation • Can generate sub-daily weather information for engineering design	• Requires arbitrary adjustment of parameters to generate climate change scenarios • Unanticipated results may arise from changing model parameters (e.g. wet-day probability can modify sunshine and temperature simulations)

atmospheric circulation patterns (such as the frequency of high-pressure systems). Weather generators are essentially random number makers that mimic the sequencing of wet and dry days from which other dependent variables such as temperature and solar radiation can be inferred. All these techniques are relatively quick and simple to implement so are particularly appealing in resource-constrained situations. However, they are only possible to apply where there are sufficient meteorological observations for model calibration.

Dynamical Downscaling Methods

RCMs are essentially fine-resolution (<25 km) climate models that simulate sub-daily information for a patch of the Earth's surface (Figure 5.3). They take GCM airflows, energy and moisture fluxes crossing the boundaries of the target domain to predict conditions within it. Because of the immense number of calculations involved, RCMs are as computationally demanding as GCMs to run. So, experiments typically cover only a few decades representing the observed (control) and changed climate. However, RCMs do provide a coherent and physically consistent expression of the climate behaviour within the domain, and can be used to investigate land surface feedbacks in addition to GHG forcing. This is a major advantage over the SD methods, which are blind to the possible influence of changing vegetation, soil moisture, albedo or snow cover on the local climate.

Comparing Downscaling Methods

Given the range of modelling options, the climate science community has understandably spent much time and resources evaluating the qualities of different downscaling techniques. Major international comparison projects have been undertaken by ACCORD, CORDEX, ENSEMBLES, NARCCAP, PRUDENCE, STARDEX and VALUE projects (see the links in Section 5.5 for further information).

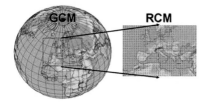

FIGURE 5.3 An RCM domain embedded in a GCM grid. *Source*: World Meteorological Organisation

From these studies it is clear that SD and RCM techniques are complementary rather than competing tools because they each have distinctive strengths and weaknesses (Table 5.1). For example, RCMs are well suited to the production of maps of regional climate change, infilling climate information in data-sparse regions (e.g. Maussion et al., 2014), or for investigating the physical processes that cause extreme events such as heavy rainfall (Figure 5.4). On the other hand, SD techniques can provide site-specific climate information as well as more 'exotic' variables such as air quality, wildfire occurrence, or storm surge height (e.g. Casanueva et al., 2014). Since the relative skill of different downscaling methods is now well understood, some are calling for research that goes beyond the classic model comparison study, focusing instead on the value added by downscaling (Fowler and Wilby, 2007; Maraun et al., 2010). In other words, is the time and effort spent creating downscaled scenarios leading to better climate risk assessment and adaptation plans? This crucial question is not an easy one to answer.

Measuring the Costs and Benefits of Downscaling (Present Climate)

Researchers are struggling to establish objective criteria for weighing the cost-benefit of regional climate downscaling. The simplest test is to determine whether differences between down-scaled information and observations (termed biases) are less than equivalent differences between observations and the host GCM. However, this depends on the choice of (i) performance metric; (ii) data used for benchmarking (with attendant measurement uncertainties and natural variability); (iii) the spatial resolution of the downscaling; and (iv) any pre-processing methods applied to climate model outputs (Haensler et al., 2011; Kanamitsu and DeHaan, 2011; Schmidli et al., 2006).

It is generally accepted that downscaling skill improves with finer resolution modelling of the present climate (Castro et al., 2005; Lee and Hong, 2014). This is because the models are more able to resolve features in complex terrain that affect local climates, especially near inland water bodies or the coastal zone (Di Luca et al., 2013). This is especially true for the intensity and diurnal variation in summer convection precipitation, which can only be simulated credibly at resolutions of 3 km or less (Prein et al., 2013). At such fine

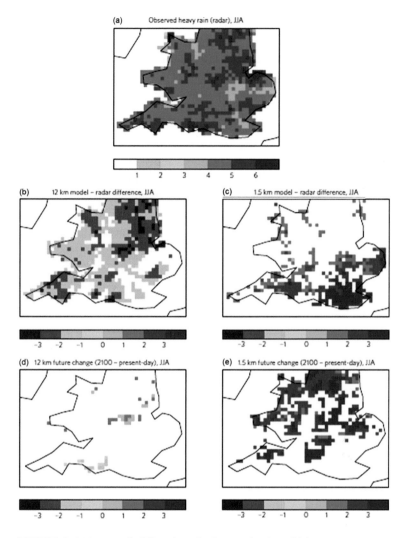

FIGURE 5.4 Heavy rainfall on hourly time scales (mm/h) in summer (June to August; JJA): (a) Observed radar; (b, c) difference between model and observed radar for 12 km and 1.5 km models, respectively; (d, e) difference between 2100 and present day for 12 km and 1.5 km models, respectively. Heavy rainfall is calculated from the mean of the top 5% of wet values. White indicates differences or future changes not significant at the 1% level compared to inter-annual variability. Radar data were bias corrected using daily rain gauge data. *Source*: Kendon et al. (2014)

scales, RCMs can represent local convection and cloud formation. Moreover, the topography is not 'smoothed' as much as within GCMs, so errors in temperature and rainfall variations due to unrealistic terrain (i.e. lapse rates) are reduced (Tselioudis et al., 2012).

Measuring the Costs and Benefits of Downscaling (Future Climate)

It is harder to establish whether value is added by downscaling under changed climate conditions. To begin, it is impossible to establish with any confidence the future boundary forcing from GCMs (which depend on unknowable GHG emissions and natural climate variability) plus local land surface feedbacks (which depend on vegetation change and aerosol emissions) that ultimately shape the regional climate (Pielke and Wilby, 2012). Furthermore, there have been very few attempts to assess model proficiency at replicating regional change, even for the past climate. One such study found that RCM skill at simulating the observed *mean* climate was not a reliable indicator of skill at capturing *change* between decades (Racherla et al., 2012). To some, this is evidence of a fundamental failure of regional climate downscaling to add value to the uncertain outlooks of future climate change provided by GCMs (Kerr, 2013). Undeterred, international consortia continue to advocate use of downscaled products for climate impact assessment and adaptation planning (e.g. Jacob et al., 2014).

Smarter Ways of Applying Downscaling

With these issues in mind, others are advocating 'smarter' use of downscaling for contemporary climate applications such as (i) improving the fidelity of seasonal climate forecasts (Yuan et al., 2013), (ii) reconstructing patchy and repairing damaged datasets, or (ii) generating plausible scenarios to test the benefits of adaptation options (Wilby et al., 2014). The first takes coarse resolution information from numerical weather predictions then downscales to the level of river basins to improve local precipitation and snowpack estimates. This in turn enhances forecasts of melt rates and volumes inflowing to reservoirs and through hydropower plants. The second group of applications add value by using archives of weather patterns to downscale and thereby repair or infill missing meteorological records (Figure 5.5). The third uses downscaling to synthesise

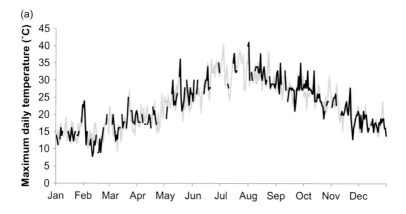

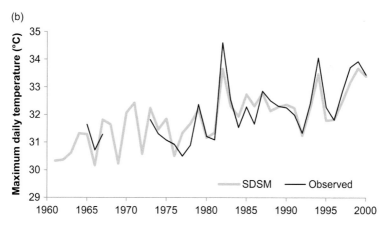

FIGURE 5.5 Statistical downscaling of maximum temperatures in Tunis shown as (a) daily series for 1965, and (b) annual mean series for 1961–2000. Missing values were estimated from GCM-scale, daily near-surface and mid-tropospheric airflows over the same period. Adapted from Wilby et al. (2014)

large sets of climate change scenarios to explore the behaviour or breaking points of systems of interest with and without adaptations. For example, the benefits of changing reservoir operating rules or raising flood defences might be compared with 'no action' under plausible scenarios of climate change (see Chapter 14).

Downscaling Applied to Real Decisions

Despite the potential advantages of downscaling there are still relatively few examples of the techniques being applied in practice – the bulk of downscaling work resides within the realm of academic research (as model comparison studies, impact assessments or scenario analysis). This may be because the accuracy of high-precision climate change scenarios is difficult to test, or even defend conceptually. Nonetheless, there are some exciting developments in downscaling, such as growing use of very high resolution (1.5 km grid spacing) models as virtual worlds within which to test process understanding. At this scale, the RCM can resolve convection processes and thereby more physically reasonable representations of severe summer storms (Figure 5.4). Seamless integration of downscaling within flood hazard forecasting and water resource allocation systems helps to reduce vulnerability and/or strengthen resilience to both present and future climate variability. In this way, downscaling supports the operational management of near-term risks. But the notion of accurate post(zip)-code, city-scale, or regional climate *predictions* for the 2050s and beyond is flawed.

Downscaling Is Only Part of the Solution

What then is the beleaguered city engineer to do about the state of the sewers in Newcastle? There are basically four options.

One is to apply the same design event as in the past. This is the cheapest solution in the short term but there is a risk of greater damages in the future if storms do become more intense.

Two is to turn to the Environment Agency (2011) *Advice for Flood and Coastal Erosion Risk Management* and look up the climate change safety factor for the 2080s. The table recommends an adjustment for heavy rainfall in the range 10–40%. Now there is a danger of spending money unnecessarily on enlarging the drainage capacity if storm intensities do not change that much. Conversely, exceptional flooding of northern UK cities during December 2015 has brought into question the adequacy of such precautionary allowances, even for the present climate.

Three is to take a more sophisticated approach by looking at the trade-off between the size of the adjustment for climate change and the cost of sewer upgrade. The design might still incorporate the most precautionary value 40% because once the roads have been dug up the additional cost of wider diameter sewers is relatively low.

Four is to integrate all the other ways of managing the surface runoff from source, such as green roofs, permeable parking areas, ponds and water retention features woven into the urban fabric.

This seems an appealing option, but consider how many organisations would have to be involved and how long it would take to implement a joined up action plan. It is also questionable whether such measures would afford much protection against the record rainfall totals seen in 2015.

What solution(s) would you choose?

5.2 GROUP EXERCISE: SENSE AND SENSIBILITY OF UKCP09

NOTES

The UK Climate Projections 2009 (UKCP09) were the fifth generation climate scenarios for the UK (Murphy et al., 2009). They marked a step change in technical complexity and approach by moving from a small set of national scenarios to partial quantification of climate model (structure and parameter) and emissions uncertainty (Figure 5.6). However, the methodology and process of scenario development have attracted both criticism (e.g. Frigg et al., 2013) and acclaim (e.g. Kelly, 2014).

The purpose of this activity is to give a deeper understanding of the development and application of probabilistic climate change scenarios, using UKCP09 as a case study. This can be achieved by viewing the suggested video clips and/or directed readings. Using the definitions of Tang and Dessai (2012), groups should record the saliency, credibility and legitimacy of the downscaled information in the accompanying worksheet.

Video clips: Review the following video clips, concentrating on evidence of saliency, credibility and legitimacy in each case. Also, record any examples of the application of the downscaled products in real decision or planning contexts:

Clip 1: Nobilis climate change downscaling capability (2 minutes)
www.youtube.com/watch?v=vTgP3o4-UCM
Clip 2: Science behind the UKCP09 projections (6 minutes)
www.youtube.com/watch?v=2zZesaRRitY
Clip 3: Rt Hon. Hilary Benn, MP announcement to Parliament (4 minutes)
www.youtube.com/watch?v=PnjHw5PeuKw

Directed readings: The suggested articles illustrate contrasting viewpoints about UKCP09 held by some boundary organisations

cont.

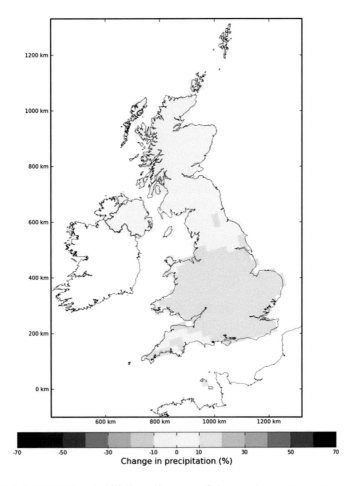

FIGURE 5.6 UKCP09 probabilistic projections of changes in mean summer rainfall totals by the 2050s under Low (B1) [above, 90th percentile] and under High (A1FI) [facing, 10th percentile] emission scenarios. *Source*: Murphy et al. (2009)

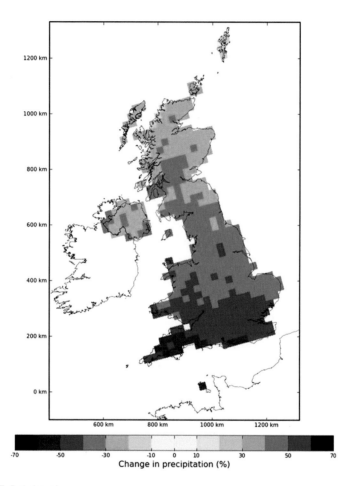

Plot Details:

Data Source: Probabilistic Land
Future Climate Change: True
Variables: precip_dmean_tmean_perc
Emissions Scenario: High
Time Period: 2040-2069

Temporal Average: JJA
Spatial Average: Grid Box 25Km
Location: -10.00, 48.00, 4.00, 61.00
Percentiles: 10.0
Probability Data Type: cdf

Change in precipitation (%)

FIGURE 5.6 (*cont.*)

(Street et al., 2009), parts of the academic community (Hulme and Dessai, 2008), and a user (Kelly, 2014). Further insight can be gained by navigating the UKCP09 portal at http://ukclimateprojec tions-ui.metoffice.gov.uk/ui/admin/login.php.

The following readings could also be set ahead of the group discussion:

Street, R.B., Steynor, A., Bowyer, P. and Humphrey, K. 2009. Delivering and using the UK climate projections 2009. *Weather*, **64**, 227–231.

Hulme, M. and Dessai, S. 2008. Negotiating future climates for public policy: a critical assessment of the development of climate scenarios for the UK. *Environmental Science and Policy*, **11**, 54–70.

Christierson, B., Vidal, J.-P. and Wade, S.D. 2012. Using UKCP09 probabilistic climate information for UK water resource planning. *Journal of Hydrology*, **424**, 48–67.

Worksheet for Sense and Sensibility of UKCP09

Outline the climate products and services on offer
Saliency (suitability for user needs)
Credibility (technical rigour)
Legitimacy (transparency of scenario design, construction, distribution)
Practicality of use for adaptation planning
Other applications of UKCP09

LEARNING OUTCOMES

- To understand the purpose and main approaches to regional climate downscaling via dynamical and statistical techniques.
- To appreciate that high-resolution (i.e. greater precision) downscaling generally improves simulations of observed *mean* climate but does not guarantee higher accuracy of climate *change* signals.

cont.

- To be aware of the different ways of assessing the 'value added' by regional climate downscaling.
- To acknowledge the large uncertainty in downscaled scenarios but recognise that these tools can still be used in smart ways to deepen understanding of regional climate change and support decision-making.

5.3 FURTHER QUESTIONS TO RESEARCH AND DISCUSS

What is the difference between accuracy and precision of regional climate change scenarios?

Why is it so difficult to measure objectively the value added by regional climate downscaling?

Give three examples of smarter use of downscaling beyond the conventional model comparison study.

5.4 FURTHER READING

Benestad, R.E., Chen, D. and Hanssen-Bauer, I. 2008. *Empirical-Statistical Downscaling*. World Scientific, Singapore and London.

Wilby, R.L. and Fowler, H.J. 2010. Regional climate downscaling. In Fung, C.F., Lopez, A. and New, M. (eds.), *Modelling the Impact of Climate Change on Water Resources*. Wiley- Blackwell, Chichester.

5.5 OTHER RESOURCES

Atmospheric Circulation Classification and Regional Downscaling (ACCORD) www.cru.uea.ac.uk/cru/projects/accord/

Bias-corrected CMIP3 and CMIP5 climate and hydrology projections http://gdo-dcp.ucllnl.org/downscaled_cmip_projections/dcpInterface.html

Coordinated Regional Climate Downscaling Experiment (CORDEX) http://wcrp-cordex.ipsl.jussieu.fr/

ENSEMBLES http://ensembles-eu.metoffice.com/index.html

North American Regional Climate Change Assessment Program (NARCCAP) www.narccap.ucar.edu/

Oxford Research Encyclopedias: downscaling climate information
http://climatescience.oxfordre.com/view/10.1093/acrefore/
9780190228620.001.0001/acrefore-9780190228620-e-27

PRECIS regional climate modelling system www.metoffice.gov.uk/
precis/

Prediction of Regional scenarios and Uncertainties for Defining
EuropeaN Climate change risks and Effects (PRUDENCE) http://
prudence.dmi.dk/

Statistical and Regional dynamical Downscaling of Extremes for
European regions (STARDEX) www.cru.uea.ac.uk/projects/stardex/

UK Climate Projections (UKCP09 and UKCP18) http://
ukclimateprojections.metoffice.gov.uk/

Validating and Integrating Downscaling Methods for Climate
Change Research (VALUE: COST Action ES1102) www.value-
cost.eu/

What Is the 'Cascade of Uncertainty' and Why Does It Matter?

TOPIC SUMMARY

Conventional approaches to climate scenario generation and risk assessment involve a step-by-step process beginning with GHG emission scenarios and ending with downscaled impacts. This 'top-down' framework causes uncertainty in socio-economic drivers and climate system responses to be propagated at each stage. Eventually, the range of uncertainty in decision-relevant metrics (e.g. crop yields or river flows) can be so wide as to confound action. Hence, the cascade of uncertainty matters because it ultimately gives way to more fruitful 'bottom-up' thinking about system vulnerabilities. Rather than attempting to predict the future this perspective accepts that uncertainty is irreducible and looks instead to test risk management strategies under a range of plausible scenarios.

BACKGROUND READING

Mitchell and Hulme (1999) were early adopters of the phrase 'cascade of uncertainty', which describes the great ambiguity in regional climate scenarios stemming from the unpredictability in the drivers and response of the climate system. Wilby and Dessai (2010) explain the difference between scenario-led and vulnerability-led climate assessment frameworks. Curry and Webster (2011) liken climate science uncertainty to a 'monster' that causes confusion, ambiguity, and even fear about the future.

6.1 SNOWBALLING UNCERTAINTIES IN REGIONAL CLIMATE IMPACTS
Risks and Opportunities from Climate Change

Tajikistan is a low income, mountainous republic situated in Central Asia north of Afghanistan. The country has beautiful scenery and welcoming people but, until recently, visitors to the capital Dushanbe could hardly fail to notice the frequent power cuts. The situation is more serious for rural communities where the winter energy crisis causes real hardship (Oxfam, 2010; Fields et al., 2012). Closing the gap between national energy demand and supply presents a major opportunity for economic development.

In 2010 electricity accounted for ~60% of total energy consumption in Tajikistan, of which 98% was produced by hydropower. Nearly three-quarters of the electricity generating infrastructure is now over 30 years old and most of the transmission network was built in the 1960s and 1970s (Figure 6.1). This ageing and deteriorating equipment is highly inefficient and vulnerable to extreme weather shocks. Hence, investors are reviewing options for upgrading existing assets and/or installing new capacity to harness the immense energy potential of major rivers such as the Naryn, Pyanj and Vakhsh (Fields et al., 2012). Plans to fix Tajikistan's electricity shortfall will have to take account of expected changes in energy demand, as well as scenarios for future tariffs, sector governance and institutional capacity. Future hydropower generation is also uncertain because it depends on the mix of energy sources (such as run of river, small hydro or dams), their physical location, design and assumed operating regimes. Ultimately, the hydropower outlook is defined by the regional climate, which drives the snowmelt-dominated river flow regimes, and long-term contributions from glacier wastage.

FIGURE 6.1 Qairokkum hydropower plant in northern Tajikistan. *Photo*: Michael Friedhoff.

Climate Risk Assessment in Practice

Within this setting, consider the practical challenges faced by a climate consultant tasked with assessing the hydropower potential of Tajikistan. His or her remit only covers energy supply, but to do this the consultant must account for plausible changes in regional temperature and precipitation, mass balance of glaciers, evaporation losses from reservoirs, as well as hazards (e.g. avalanches, mud flows, flash floods, heatwaves) that might interrupt operations. The indirect impacts of land degradation on sediment supply and loss of reservoir storage further complicate the task. Even more challenging, these analyses have to be completed within several months and on the back of very limited data. Those hydrometeorological series that can be found may have ended decades ago or require digitisation from paper records. The highly varied terrain (Figure 6.2) means that a sparse network of observing stations barely characterises the present climate let alone a baseline for detecting future changes.

Identifying Sources of Uncertainty

The consultant would think about the available options. It will be recalled that one of the most significant sources of uncertainty affecting regional climate change projections originates from the host general circulation models (GCMs). The choice of downscaling method (Chapter 5) and impact model are also important but probably of secondary concern. Moreover, the required spatial and temporal resolution of the scenarios, the need to simulate transient climate change and river flow impacts, and the limited amount of observational data further constrain what is feasible (Wilby et al., 2009). The climate change practitioner reviews the range of tools available and recognises that there could be very different time and

FIGURE 6.2 Tajikistan is one of the most climate vulnerable nations in Central Asia, not least because the republic has high dependency on hydropower fed by snow and glacier melt. *Photo:* Alex Harvey.

resource implications depending on the choice made. Having referred to the IPCC and United Nations' online guidance material (see Section 6.5) a decision is possibly made to investigate the effect of five factors on Tajikistan's future hydropower production:

1. GREENHOUSE GAS EMISSIONS

The pace of regional climate change and implied hydropower potential is ultimately governed by global GHG emissions. These depend upon assumed patterns of socio-economic development, in addition to demographic and technological change. Hence, very different emission pathways and radiative forcing of the climate can emerge. Until the IPCC Fifth Assessment Report these futures were described as Special Report on Emissions Scenarios (SRES) (Nakicenovic and Swart, 2000) (Figure 6.3). Two key 'axes' of change were represented: (i) environmental versus economic and (ii) globalisation versus regionalisation of markets and cultures. From these, four families of scenarios emerge: A1 (global, economic); A2 (regional, economic); B1 (global, environmental); B2 (regional, environmental).

The SRES scenarios were subsequently replaced by Representative Concentration Pathways (RCPs), which recognise that similar atmospheric concentrations of CO_2 can arise from different combinations of socio-economic, demographic, land use and technological change (Moss et al., 2010). This has the advantage of mixing and matching different climate forcing scenarios against different socio-economic pathways (with implied scope for climate mitigation and adaptation). A group of scenarios (i.e. RCP2.6, RCP4.5, RCP6.0 and RCP8.5) charts the change in radiative forcing by 2100 relative to pre-industrial conditions of 2.6, 4.5, 6.0 and 8.5 W/m^2, respectively. The most extreme (RCP8.5) imagines a global population of 12 billion by the end of the twenty-first century, with a low rate of technology development leading to highly intensive energy consumption (strong reliance on coal) and GHG emissions. Overall, RCP8.5 has less global mean warming between 2035 and 2080 than the most extreme SRES scenario (A1FI), but greater median temperature changes beyond 2080 (Figure 6.3).

2. GLOBAL CLIMATE MODELS

Numerous studies have shown that the GCM ensemble critically determines regional climate change scenarios over all time horizons. In comparison, uncertainty due to choice of GHG emissions is smaller in the near term (2020s) than at the end of century (2080s).

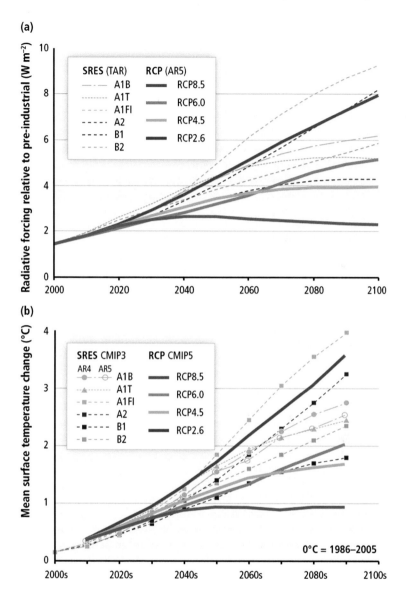

FIGURE 6.3 Projected (a) radiative forcing (W/m²) and (b) mean temperature change over the twenty-first century using the Special Report on Emissions Scenarios (SRES) of the Third Assessment Report (TAR) and Representative Concentration Pathway (RCP) scenarios of the Fifth Assessment Report (AR5). *Source*: Burkett et al. (2014), Figure 1.4a

According to IPCC (2013a:20): "*global surface temperature change for the end of the 21st century is* likely[1] *to exceed 1.5°C relative to 1850 to 1900 for all RCP scenarios except RCP2.6. It is* likely *to exceed 2°C for RCP6.0 and RCP8.5, and* more likely than not[2] *to exceed 2°C for RCP4.5. Warming will continue beyond 2100 under all RCP scenarios except RCP2.6. Warming will continue to exhibit inter-annual-to-decadal variability and will not be regionally uniform.*" The bars around each RCP pathway in Figure 6.4 show the extent of climate model uncertainty within the CMIP5 ensemble for the global mean. For global mean temperature, the likely range for RCP2.6 is 0.3–1.7 °C, and for RCP8.5 it is 2.6–4.8 °C by 2081–2100. Climate model uncertainty is much greater for regional precipitation. For example, by 2080–2099 the range in projected annual precipitation change across the Naryn River basin, Central Asia, spans −25% to +40% (Figure 6.5).

3. NATURAL CLIMATE VARIABILITY
Part of the uncertainty shown in Figure 6.5 is due to climate variability within the ensemble of model simulations. This is evident in the spread of results (realisations) produced by a single climate model under a given RCP pathway. For precipitation changes over the River Naryn, climate variability can spread the scenario across +25% to +40% in one model or by −10% to +0% in another by the end of the twenty-first century. Likewise, Deser et al. (2014) report near equal odds of summer drying or wetting across the continental United States over the next 50 years. This modelled climate variability is simply due to slightly different atmospheric conditions used at the start of each experiment (see Figure 3.5). Hawkins et al. (2016) showed that even greater regional climate uncertainty can emerge if different initial ocean states are used instead of different atmospheric conditions. These wide ranges of uncertainty may still be conservative because climate models are known to underestimate the natural variability in important modes of variability such as the North Atlantic Oscillation (Scaife et al., 2009).

4. DOWNSCALING METHOD(S)
Downscaling technique(s) and method(s) of calibration further compound the uncertainty from emissions, climate model and natural variability. In practice, the choice of downscaling method depends very much on the intended application, amount of time and data

---

[1] *Likely* is assessed as 66–100% probability
[2] *More likely than not* is assessed as >50% probability

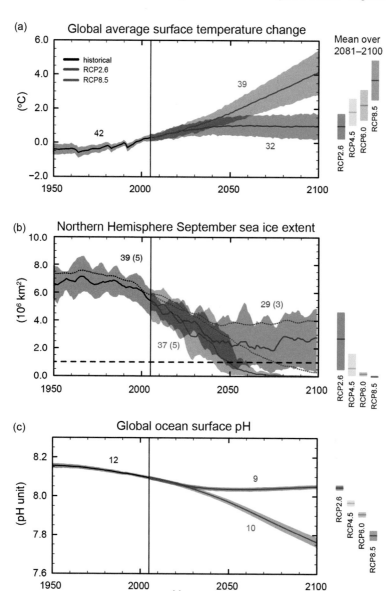

FIGURE 6.4 CMIP5 multi-model simulated time series from 1950 to 2100 for change in global annual mean surface temperature relative to 1986–2005. Time series of projections and a measure of uncertainty (shading) are shown for scenarios RCP2.6 (blue) and RCP8.5 (red). Black (grey shading) is the modelled historical evolution using historical reconstructed forcings. The mean and associated uncertainties averaged over 2081–2100 are given for all RCP scenarios as coloured vertical bars. The numbers of CMIP5 models used to calculate the multi-model mean is indicated. *Source*: IPCC (2013a), Figure SPM.7

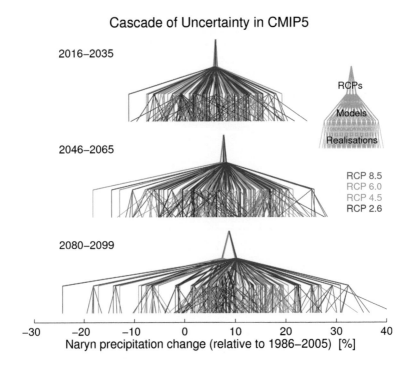

FIGURE 6.5 Annual precipitation changes projected by the CMIP5 ensemble for the River Naryn basin, Central Asia (70–80° E, 40–45° N). The three levels of each pyramid illustrate uncertainty due to the choice of Representative Concentration Pathway (RCP), GCM and climate variability (shown by multiple lines for some GCMs). The intersection on the top row for each time period is the multi-RCP, multi-model, multi-run mean precipitation change. Note that this plot does not show the uncertainty added by different downscaling methods. Source: Ed Hawkins and Wilby et al. (2014) © 2014 Inter-Research

available, and on the climate variables of most relevance to the study. In general, there is no universally supreme downscaling method: some are more skilful in some times and places and for some indices than others (Figure 6.6). Therefore, a pragmatic decision is typically made based on resource implications.

A consultant interested in hydroclimatic variability and change at the scale of individual river basins would want a downscaling tool that can reproduce seasonal to inter-annual variations in temperature and precipitation (Figure 6.7). However, site-by-site calibration

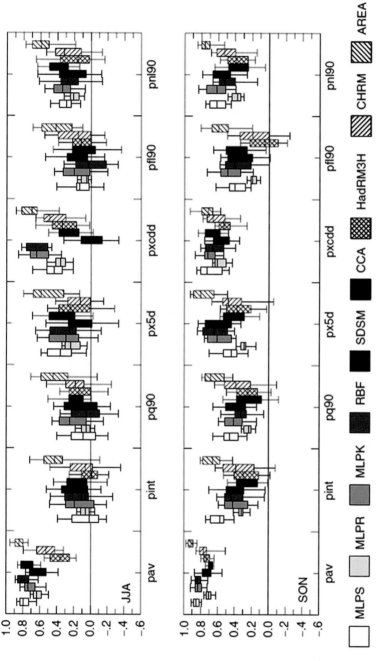

FIGURE 6.6 A comparison of the skill of nine downscaling methods using correlation with observed precipitation indices (pav, pint, pq60, etc.) for summer (JJA) and autumn (SON) in southeast England. Bars show the 5th to 95th percentile range, boxes the interquartile ranges, and horizontal lines the median correlation (skill) across all sites assessed. The metric codes are: pav (average precipitation on all days); pint (average precipitation on days with >1 mm); pq90 (90th percentile of precipitation on days with >1 mm); px5d (maximum precipitation from any five consecutive days); pxcdd (maximum number of consecutive days with <1 mm); pfl90 (fraction of total precipitation from events > long-term 90th percentile); pnl90 (number of events > long-term 90th percentile). *Source:* Haylock et al. (2006)

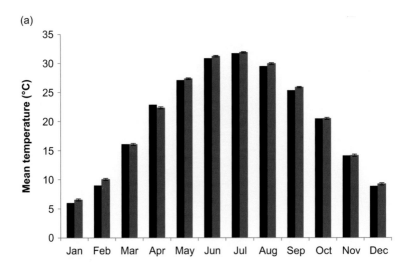

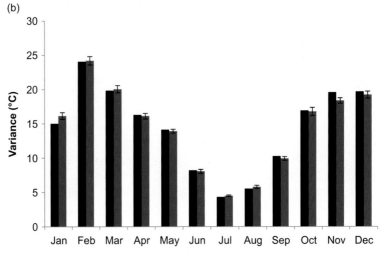

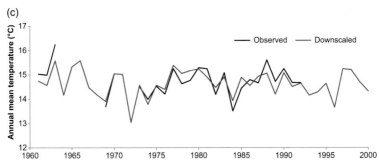

FIGURE 6.7 Observed (black) and downscaled (red) monthly (a) mean, (b) variance and (c) annual mean air temperatures at Khujand, Tajikistan. *Source*: Wilby (2010b)

may be too cumbersome for a rapid assessment, or the necessary predictor variables might not have been archived (see Crawford et al., 2007; Wilby and Wigley, 2000). Time must also be spent checking the statistical adequacy of all downscaling models (Estrada et al., 2013). Therefore, many climate impact assessments default to much simpler techniques such as climate change factors (Diaz-Nieto and Wilby, 2005) or bias correction methods (Pierce et al., 2015). These scale observed meteorological data using the changes at the nearest GCM grid point. Although quick to apply, the methods are known to underestimate the persistence and severity of extreme weather events.

5. IMPACT MODELLING

By now the consultant would be forgiven for feeling somewhat overwhelmed by the proliferating uncertainties, yet it would be naive to assume that the impact model is somehow different. For example, Lutz et al. (2013) showed that temperature and precipitation signals derived from the CMIP3 and CMIP5 ensembles yield a 30–45% reduction in glacier mass balance for the Amu Darya and Syr Darya river basins by 2050. This translates into similar declines in glacierised area over the same period. However, uncertainty in the glacier model parameters spreads this range to about 25–55%. More fundamental differences in outlook emerge if mass balance and dynamic equilibrium glacier models are compared. The former anticipates progressive declines in melt and river flow as the glaciers waste at lower elevations (Lutz et al., 2013); the latter foresees an initial increase in annual mean discharge to year 2060 and thereafter a decline (Kure et al., 2013). Hence, regardless of the uncertainty in downscaled climate change scenarios, the sign of the assessed change (to reservoir inflows) can be positive or negative over the investment time horizon depending on the choice of glacier model (Figure 6.8). The inflow scenarios still have to be fed into yet another model to convert reservoir levels into hydropower production...

Other Practical Considerations

The above step-by-step assessment is technically demanding, but the consultant would still have to overcome other logistical

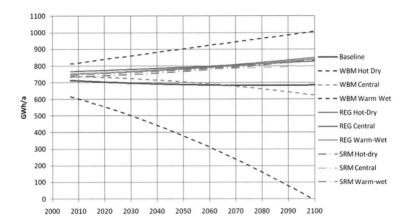

FIGURE 6.8 Simulations of future annual energy production (GWh/a) at Qairokkum hydropower plant (Figure 6.1). Scenarios were based on three hydrological (impact) models (WBM, REG, SRM) and three climate change scenarios (Hot-dry, Central, Warm-wet) which span the range of GCMs used in the IPCC Fourth Assessment Report (CMIP3 runs). The slight reduction in baseline energy production is due to reservoir sedimentation and associated loss of storage. *Source*: European Bank for Reconstruction and Development (2011)

challenges. For instance, daily temperature and precipitation records are needed for glacier and river flow modelling. Unfortunately, such data are costly to obtain or are in short supply in Central Asia (see Figure 1.3). Fortunately, the situation is improving thanks to recent initiatives to strengthen hydrometeorological networks and forecasting systems. There are also portals where daily meteorological data can be freely accessed (e.g. KNMI Climate Explorer[3]). Even so, public data can be in unfriendly formats or require further quality assurance. In many situations daily records are unlikely to be available for recent years (Figure 1.4), further limiting the possibility of applying any statistical downscaling. Short-lived or broken hydrometeorological records (Figure 5.5) also confound efforts to weigh up the realism of the alternative melt scenarios. There might not even be recent river flow records to judge if the direction of travel in annual flows is up or down (e.g. Chevallier et al., 2012).

---

[3] http://climexp.knmi.nl/selectdailyseries.cgi?id=someone@somewhere

Limitations of 'Top-Down' Frameworks

Having thought through steps 1 to 5, the consultant might conclude that it is simply not practicable (or even helpful) to quantify all uncertainty components affecting downscaled scenarios. This would involve characterising uncertainty contributed by a host of factors including: observations (which depends on period of record used to calibrate the models); downscaling domain and predictor set; downscaling technique; emission scenarios; ensemble(s) of host GCMs; natural climate variability; then finally a suite of impact models. Even then, the true range of uncertainty, as wide as it already is, would be underestimated because of hidden parameters as elemental as the assumed temperature lapse rate over ice (e.g. Dobler et al., 2012). Nonetheless, the vast majority of climate risk assessments to date apply this so-called 'top-down' framework in which the scenario generation is the entry point and the evaluation of impacts and actions the last step. (See, for example, the meta-analysis of downscaling studies by Wilby and Dawson (2013).) The cascade of uncertainty could paralyse the whole effort to adapt the vulnerable infrastructure.

Vulnerability-Based Approaches

However, the consultant would realise that there are more effective ways of applying the same tools and climate risk information by using decision-analytic (Brown et al., 2011) or scenario-neutral approaches (Prudhomme et al., 2010). These invert the framework (now 'bottom-up') and start by asking what decisions and adaptation options are socially, economically and technically feasible and then how would these perform under stressful climate conditions? In this form of analysis, regional climate downscaling is used to probe system vulnerabilities across a range of plausible scenarios (see Wilby and Dessai, 2010).

All the above practical concerns about choice of scenario tool, access to data, impact model calibration and testing still apply but they no longer take centre stage. Instead, the emphasis is on discovering critical vulnerabilities and tipping points in the system, as well as technical options that are robust to uncertainty (EBRD, 2015). The consultant advises the client that the first investment should be to strengthen the network of hydromet stations. The logic is that

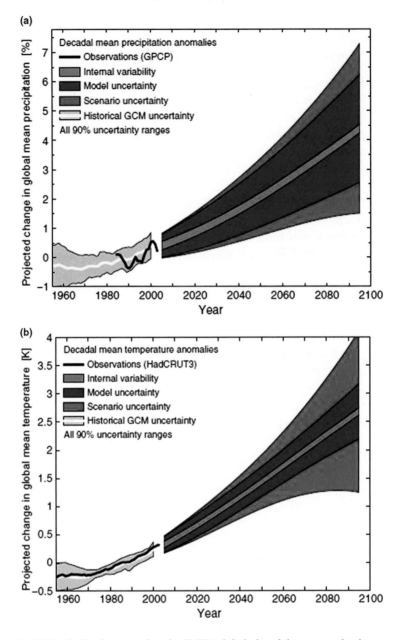

FIGURE 6.9 Total uncertainty in CMIP3 global, decadal mean projections for the twenty-first century, separated into internal variability (orange), climate model uncertainty (blue) and emissions scenario uncertainty (green). Grey regions show the uncertainty in the twentieth century runs of the same GCMs, with the mean in white. Black lines show an estimate of the observed historical changes. (a) Precipitation, with observations from GPCP v2.1. (b) Temperature, with observations from HadCRUT3. Anomalies are calculated relative to the 1971–2000 mean for temperature and 1979–2000 mean for precipitation. *Source*: Hawkins and Sutton (2010)

without basic observations of environmental change there is very little scope for any form of climate assessment, whether 'top-down' or 'bottom-up'.

6.2 GROUP EXERCISE: VAKHSH CASCADE OF UNCERTAINTY

NOTES

This chapter shows how uncertainties about future GHG emissions and global climate responses are magnified at the scale of regional impact assessment and adaptation planning. Moreover, multi-model studies show that natural climate variability and climate model uncertainty dominate in the short (2020s) to medium term (2050s), but the emissions pathway is the major component of uncertainty in the long term (2080s) (Figure 6.9). Downscaling and impact model uncertainty are generally regarded as secondary concerns (Dobler et al., 2012; Wilby and Harris, 2006).

This activity makes direct use of climate model outputs via the user-friendly KNMI Climate Explorer. The large uncertainty in regional temperature and precipitation projections (for Tajikistan) will soon become apparent. The sobering message is that these plausible ranges of future climate, wide as they are, do not reflect the true extent of uncertainty even if downscaling and impact modelling are included. This should prompt discussions about smarter ways of using climate scenarios to aid decision-making.

Worksheet for Vakhsh Cascade of Uncertainty

The objective of this exercise is to assess the range of uncertainty in regional climate change scenarios that might inform energy projections and safety at a hydropower plant in Tajikistan (e.g. Nurek scheme at 38°40′ N, 69°46′ E). This involves making choices about (a) the spatial scale (*Region*) of analysis; (b) relevant time period (*Season*); (c) GCM ensemble(s) (*Dataset*); (d) impact-relevant metric(s) (*Variables*); (e) type of graphical presentation(s) (*Output*); (f) emissions pathway(s) (*Scenario*); (g) time horizon (*Reference and Future Periods*); and (h) point(s) on the distribution assessed (*Mean/percentile*).

Intended application	Scenario			
	S1	S2	S3	S4
(a) Region				
(b) Season				
(c) Dataset				
(d) Variable				
(e) Output				
(f) Scenario				
(g) Reference and Future Periods				
(h) Mean or Percentile				
(i) Scenario range				

Use the KNMI Climate Explorer (http://climexp.knmi.nl/plot_atlas_form.py) to develop a set of climate change scenarios (labelled here as S1 to S4) for relevant variables at the appropriate scale of analysis. Record your choices in steps (a) to (h) above.

Based on these scenarios how would you expect hydropower supply to change across Tajikistan over coming decades?	What other sources of information would help to inform long-term investment decisions in the region's energy sector?

LEARNING OUTCOMES

- To appreciate that uncertainty in regional climate change scenarios arises from ambiguity about a host of factors including future socio-economic development, greenhouse gas emissions, global and downscaling model representations of regional climate, natural climate variability, impact model type and parameters, and observational data.
- To understand that the relative significance of these uncertainty components changes in time with natural variability dominating temperature and precipitation

cont.

projections until the 2040s and emissions uncertainty beyond this time.
- To be aware that regional climate change scenarios can be used in different ways depending on whether 'top-down' (uncertainty-focused) or 'bottom-up' (vulnerability-focused) frameworks are applied.

6.3 FURTHER QUESTIONS TO RESEARCH AND DISCUSS

If you were performing a climate risk assessment for hydropower in Central Asia, what would be the most significant sources of uncertainty to consider for the 2050s?

Explain the difference between a 'top-down' and a 'bottom-up' framework for climate analysis and how downscaling might be implemented in each case.

Compare the practical considerations that would inform the choice of downscaling method for a climate risk assessment in (i) Canada and (ii) Zimbabwe?

6.4 FURTHER READING

Fowler, H.J., Blenkinsop, S. and Tebaldi, C. 2007. Linking climate change modelling to impact studies: recent advances in downscaling techniques for hydrological modelling. *International Journal of Climatology*, **27**, 1547–1578.

Van Vuuren, D.P. and Carter, T. 2014. Climate and socio-economic scenarios for climate change research and assessment: reconciling the new with the old. *Climatic Change*, **122**, 415–429.

6.5 OTHER RESOURCES

Climate Investment Funds – Tajikistan www-cif.climateinvestmentfunds.org/country/tajikistan

IPCC Task Group on Data and Scenario Support for Impact and Climate Assessment (TGICA): *General Guidelines on the Use of Scenario Data for Climate Impact and Adaptation Assessment* www.ipcc-data.org/guidelines/TGICA_guidance_sdciaa_v2_final.pdf

KNMI Climate Change Atlas http://climexp.knmi.nl/plot_atlas_
form.py

United Nations compendium on methods and tools to evaluate
impacts of, and vulnerability and adaptation to, climate change
http://unfccc.int/adaptation/nairobi_work_programme/
knowledge_resources_and_publications/items/5402.php

7

What Shapes Climate Vulnerability?

TOPIC SUMMARY

Vulnerability is defined here as the predisposition to be adversely affected by climate variability and change. This depends on the level of *exposure* of a receptor to climate hazards, the *sensitivity* of that receptor to harm, and the *capacity* of the receptor to adapt. Receptors may be individuals, households or communities. They may also be natural systems, physical assets, sectors, cities or nations. Exposure, sensitivity and adaptive capacity can change rapidly during severe weather, or more gradually over years to decades. Hence, climate vulnerability is multi-faceted, geographically and socially differentiated. Survey data and socio-economic metrics are often combined to track changes in vulnerability over space and time. These indicators help to identify the most important drivers of vulnerability and thereby target resources for adaptation. Extreme events such as Super Typhoon Haiyan reveal the extent to which poor households in developing regions are particularly vulnerable.

BACKGROUND READING

Kelly and Adger (2000) introduce the concepts of social vulnerability and adaptation then illustrate these ideas using a case study of tropical storm impacts in coastal areas of the Red River Delta, Vietnam. Füssel and Klein (2006) offer a historical perspective on the development of thinking behind the IPCC approach to climate change vulnerability assessment.

7.1 EXPOSURE, SENSITIVITY AND ADAPTIVE CAPACITY OF PEOPLE AND PLACES

A Climate Catastrophe

Super Typhoon Haiyan (known locally as Yolanda) made landfall on 8 November 2013 in Eastern Samar, a province of the Philippines. With maximum wind speeds exceeding 270 km/hr reported by the Hong Kong Observatory, this typhoon was amongst the most powerful ever recorded. Eye-witness accounts described a 5–6 m high storm surge that struck downtown Tacloban and led to the majority of fatalities (Lagmay et al., 2015). It is estimated that the storm claimed up to 7300 lives, injured nearly 29,000, and displaced 4.1 million after their homes were destroyed or damaged (Overseas Development Institute [ODI], 2015). More than 3000 schools and other critical infrastructure were partially or totally demolished. Such was the scale of devastation caused by Haiyan that the Germanwatch Climate Risk Index ranked the Philippines as the country most affected by climate change in 2013.

Changing Cyclones

Opinion is divided about whether the properties (i.e. severity, frequency and duration) of intense tropical cyclones (typhoons) are changing. Trend detection is hampered by large year-to-year variations in storm frequency and intensity, as well as by short observational records. Satellite data covering the period 1982–2009 suggest that the lifetime maximum intensity of tropical cyclones has on average increased globally, but actually fell by 2 m/s per decade in the western North Pacific (Kossin et al., 2013). Others shy away from linear trend analysis and instead point to fluctuations in cyclone energy linked to ENSO. Although the physical causes are uncertain, it appears that during El Niño years the cyclones tend to be more intense and longer lived than in La Niña years (Camargo and Sobel, 2005).

One very high resolution surge-wave tide model showed the extent to which higher Pacific sea surface temperatures (SSTs) due to anthropogenic climate change had increased the severity of Haiyan relative to simulated 'natural' conditions (Takayabu et al., 2015). The most extreme model experiment suggests that the height of the storm surge could have been amplified by ~20% due to historical GHG emissions. However, such results must be regarded with caution because predicted increases in the frequency of high-intensity storms (and associated damages) are known to depend on the choice of climate model (Mendelsohn et al., 2012). Overall, there is an expectation that the intensity (i.e. wind speeds and heavy rainfall) of tropical cyclones will increase by 2100 (IPCC, 2012).

Socio-economic Status and Vulnerability

That extreme weather events disproportionately impact the poor is not in doubt. At the global scale, the 10 *deadliest* cyclones since 1900 have all struck in developing countries (Bangladesh, China, Myanmar and India) whereas 9 out of 10 of the *costliest* in terms of the economic damages have all been in the United States (Table 7.1). The tenth most costly tropical cyclone was, in fact, Haiyan at US$10 billion (2013 prices), but according to EM-DAT only 7% of this loss was insured (compared with nearly half of the losses caused by Hurricane Katrina in 2005). Unfortunately, there is a strong overlap between the places expected to experience most rapid climate change and the least developed regions (Figure 7.1). Hence, sub-Saharan Africa is widely regarded as one of the areas most vulnerable to climate change.

National statistics for weather-related losses can be misleading because of the greater amount of exposed assets and wealth of middle income and developed countries. The true cost in terms of loss of livelihoods and productive assets is more telling. After Haiyan hit the Philippines it was apparent that the majority of victims were dependent on agriculture- and fishing-based livelihoods that require access to the coast (ODI, 2015). According to one post-disaster survey, 72% of victims self-evaluated themselves as poor, compared with 52% for the rest of the country (Social Weather Station, 2014). In 2013 the Government declared a 'no-dwell zone' within 40 m of the coast, yet some of the poorest fishers had no choice but to rebuild homes in high-risk lowland areas close to the source of their livelihood.

Preparing for Hazards

Many lives were saved thanks to greater preparedness and mass evacuations ahead of Haiyan's landfall. Nationwide it is estimated that 99.97% of the population received early warnings about the approaching typhoon. Nonetheless, interviews with survivors suggest

TABLE 7.1 The 10 deadliest and most costly tropical cyclones since 1900. *Data source: EM-DAT: The OFDA/CRED International Disaster Database – www.emdat.be, Université Catholique de Louvain, Brussels (Belgium)*

Most deadly					Most costly				
Date	Cyclone	Country	Fatalities	Cost*	Date	Cyclone	Country	Fatalities	Cost*
19/11/1970	Great Bhola	Bangladesh	300,000	<0.1	28/8/2005	Katrina	USA	1833	125
29/4/1991	Chittagong	Bangladesh	138,866	1.8	28/10/2012	Sandy	USA	54	50
20/5/2008	Nargis	Myanmar	138,366	4	12/9/2008	Ike	USA	82	30
27/7/1922	Swatou	China	100,000	–	24/8/1992	Andrew	USA	44	27
10/1942	Sundarbans	Bangladesh	61,000	–	15/9/2004	Ivan	USA	52	18
1935	Bay of Bengal	India	60,000	–	23/9/2005	Rita	USA	10	16
8/1912	Wenchou	China	50,000	–	13/8/2004	Charley	USA	10	16
14/10/1942	Orissa	India	40,000	–	24/10/2005	Wilma	USA	4	14
11/5/1965	Barisal	Bangladesh	36,000	<0.1	5/9/2004	Frances	USA	47	10
28/5/1963	Chittagong	Bangladesh	22,000	<0.1	8/11/2013	Haiyan	Philippines	7354	10

* Damage value in US$ billion at the time of the event

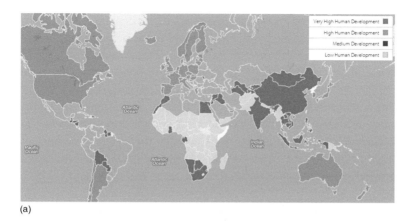

(a)

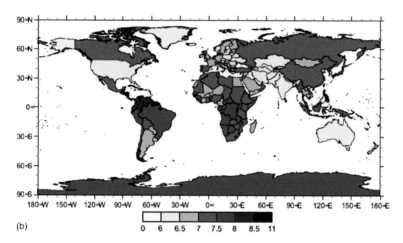

(b)

FIGURE 7.1 (a) The United Nations Development Programme (2014) Human Development Index (HDI) for 2013 compared with (b) the composite climate change index (CCI) of Baettig et al. (2007) [lower panel]. The HDI is based on four metrics: life expectancy at birth, mean years of schooling, expected years of schooling and gross national income per capita. The CCI is based on four indicators describing expected changes in annual temperatures, annual precipitation, extreme temperature events and extreme precipitation events. *Source*: (a) UNDP http://hdr.undp.org/sites/default/files/map_1.png

that some residents in affected areas did not fully comprehend the meaning of 'storm surge' (ODI, 2015). These 'bulges' in normal astronomical tide levels are caused by the passage of low-pressure systems which also produce very high wind speeds and waves.

Longer-term initiatives by the Government to reduce exposure to hazards and build adaptive capacity include upgrading key infrastructure and municipal planning of settlements away from lowland areas. Detailed surge inundation maps help to identify less exposed sites for critical infrastructure. These maps also inform levels of protection needed for structures in flood zones, and raise public awareness about hazards (e.g. Tablazon et al., 2015). The authorities further recognise the natural defence to storm surge afforded by mangroves and are re-establishing forests in the Eastern Visayas. Such efforts must counter the legacy of decades of rapid urban growth, often comprising poorly constructed (untitled) properties on unstable slopes, coastal areas and floodplains where land is cheaper.

Exposure, Sensitivity and Adaptive Capacity

The uneven impact of Haiyan is consistent with the view that [climate] vulnerability is geographically and socially differentiated, and processes that mediate the outcomes of hazard events operate at the local scale (Brooks et al., 2005:162). The IPCC (2001) broke vulnerability down into three components: exposure, sensitivity and adaptive capacity. Applying this typology to the Philippines it is evident that the country is one of the most disaster prone in the world because of exposure to about 20 typhoons annually. However, exposure also depends on choices about where to build homes and the types of livelihoods adopted. Sensitivity to typhoons is increased by factors such as poor construction and degradation of mangrove forests. Adaptive capacity is enhanced through early warning systems (via radio, television and phones), contingency planning and disaster risk management strategies. It also depends on many other factors including household income and livelihood. Poverty increases climate vulnerability by raising exposure and sensitivity to hazards, whilst presenting obstacles to adaptation. Consequently, the poor are disproportionately impacted by climate hazards, so it follows that poverty reduction is a way of improving resilience to climate variability and change (Adger et al., 2003).

Units of Vulnerability Assessment

Climate vulnerability may be assessed using various receptors or 'analytical lenses' (Leichenko et al., 2010). In the case of the Philippines, vulnerability to storm surges might be determined at individual, household, social group, city, region or sectoral levels. These receptors are subject to different social and geographical vulnerabilities. In other words, climate vulnerability is a function of who and where you are in relation to existing and emerging hazards. Note that climate vulnerability can also be determined for environmental systems (e.g. key biota, conservation sites, biomes, landscape units), natural resources (e.g. water, food and energy supplies) and infrastructure (e.g. electricity generation and distribution networks, water treatment works, transport and communication systems). Underlying trends in non-climatic pressures such as an aging and/or growing population may increase long-term vulnerability to climate variability and change. For example, the rapid growth of cities in South Asia that are already experiencing lethal heatwaves is creating even greater concentrations of people exposed to projected rises in extreme air temperatures (Murari et al., 2015).

Vulnerable Households in Rural/Shoreline Communities

The household is a meaningful unit for studying control over assets and livelihood strategies and therefore a firm basis for determining climate vulnerability (e.g. Harvey et al., 2014; Heltberg et al., 2009). The household livelihood vulnerability index (LVI) of Hahn et al. (2009) is amongst the best known. The LVI captures all three aspects of vulnerability using indices for (i) natural disasters and climate variability (exposure); (ii) health, food and water resources (sensitivity); (iii) socio-demographic profile, livelihood and social networks (adaptive capacity). In the original study in Mozambique, the researchers first consulted village leaders then randomly selected heads of household for interview. Survey data were next weighted and aggregated to district level to enable comparison of communities. Although this LVI was devised for agricultural and fishing communities in Mozambique, the template is readily applied to other situations. For instance, the natural disaster sub-component in Table 7.2 could be augmented with questions about the percentage of houses within 40 m of the coast, presence or absence of mangrove forests, or data on regional rates of sea level rise to better reflect the hazards of the Philippines. Likewise, questions about livelihoods could cover income from tourism.

TABLE 7.2 Components of a livelihood vulnerability index.
Adapted from: Hahn et al. (2009)

Socio-demographic profile (adaptive capacity)	Ratio of the population under 15 and over 65 years of age to the population between 19 and 64 years of age
	Percentage of households where the primary adult is female
	Percentage of households where the head of the household reports that they have attended 0 years of school
	Percentage of households that have at least 1 orphan living in their home
Livelihood (adaptive capacity)	Percentage of households that report at least 1 family member who works outside of the community for their primary work activity
	Percentage of households that report only agriculture as a source of income
	The inverse of (the number of agricultural livelihood activities + 1) reported by a household
Health (sensitivity)	Average time it takes the households to get to the nearest health facility
	Percentage of households that report at least 1 family member with chronic illness
	Percentage of households that report at least 1 family member who had to miss school or work due to illness in the last 2 weeks
	Months reported exposure to malaria. Owning at least one bednet indicator (have bednet = 0.5, no bednet = 1)
Social networks (adaptive capacity)	Give–receive ratio of (the number of types of help received by a household in the past month + 1) to (the number of types of help given

cont.

TABLE 7.2 (cont.)	
	by a household to someone else in the past month + 1)
	Ratio of number of households borrowing money in the past month to number of households lending money in the past month
	Percentage of households that reported that they have not asked their local government for any assistance in the past 12 months
Food (sensitivity)	Percentage of households that get their food primarily from their personal farms
	Average number of months households struggle to obtain food for their family
	Crop diversity index expressed as the inverse of (the number of crops grown by a household + 1)
	Percentage of households that do not save crops from each harvest
	Percentage of households that do not have seeds from year to year
Water (sensitivity)	Percentage of households that report having heard about conflicts over water in their community
	Percentage of households that report a creek, river, lake, pool or hole as their primary water source
	Average time it takes the households to travel to their primary water source
	Percentage of households that report that water is not available at their primary water source every day
	The inverse of (the average number of litres of water stored by each household + 1)

cont.

TABLE 7.2 (cont.)	
Natural disasters and climate variability (exposure)	Total number of floods, droughts and cyclones that were reported by households in the past 6 years
	Percentage of households that did not receive a warning about the most severe flood, drought and cyclone event in the past 6 years
	Percentage of households that reported either an injury to or death of one of their family members as a result of the most severe flood, drought, or cyclone in the past 6 years
	Standard deviation of the average daily maximum temperature by month between 1998 and 2003 averaged for each province
	Standard deviation of the average daily minimum temperature by month between 1998 and 2003 averaged for each province
	Standard deviation of the average monthly precipitation between 1998 and 2003 averaged for each province

Vulnerable Places in Cities

The city is another favoured unit for climate vulnerability assessment, not least because urban areas are now home to more than half of humankind (see Chapter 10). Vulnerability of urban populations can be determined both *within* and *between* cities. For example, Porio (2011) undertook a survey of 300 households situated in flood-prone areas of Metro Manila in the Philippines and found that most had low incomes, lived in slum/squatter settlements in substandard housing and did not have adequate access to potable water, electricity, health, sewerage or sanitation facilities. This coincidence of the poorest households (in Kalookan, Malabon Navotas and Tguig) (Figure 7.2) with parts of the city most at risk to flooding has the potential to magnify climate change impacts. Female-headed households are particularly vulnerable

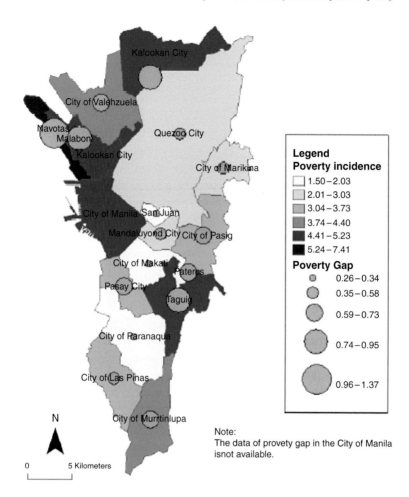

FIGURE 7.2 Incidence of poverty in Metro Manila. Poverty was defined as the percentage of families/individuals with per capita daily income less than US$1 (in 2003); the poverty gap indicates the percentage shortfall from the poverty threshold. *Source*: Map 4 in Nakamura (2009)

as they incur higher costs and a greater number of absences from school/lost workdays after flooding (Porio, 2014). Moreover, the women in these households also bear a heavier burden of post-flood cleaning of homes and care for the sick. Rufat et al. (2015) observe that social vulnerability varies during the course of a flood disaster: children are more vulnerable at flood outset due to lower awareness and preparedness; men adopt more risk-taking behaviours and are involved in rescue/emergency operations during the flood; children, the elderly and disabled may experience greater difficulty swimming or seeking shelter; women, single parents and elderly may be disadvantaged after the event when resources are scarce.

Most Vulnerable Cities

The large number of fatalities and scale of destruction caused by Haiyan was a wake-up call for the Government of the Philippines (ODI, 2015). With a population of over 20 million residing in the western North Pacific typhoon corridor, and the high concentration of poor residents in low-lying flood-prone areas, Manila has been identified as the most vulnerable city in the world to sea level rise and storm surges. Brecht et al. (2012) estimate that a 10% increase in the intensity of a 1-in-100-year typhoon combined with sea level rise and population growth to year 2100 could expose an additional 3.4 million people to storm surge. In fact, Manila would account for a quarter of the total increase across all developing country urban populations exposed to future storm surges. The next five most vulnerable cities (ranked by size of population potentially affected) are all located in Asia: Karachi (Pakistan), Jakarta (Indonesia), Khulna (Bangladesh), Kolkata (India) and Bangkok (Thailand). These cities also contribute large proportions of their national economic activity. Hence, it has been estimated that a weather disaster affecting Manila could potentially disrupt over 25% of the Philippines' gross domestic product (Hochrainer and Mechler, 2011).

Indicators of National Vulnerability

Climate vulnerability is also quantified at national scales to prioritise resources for adaptation and to monitor progress/evaluate policy effectiveness. For example, Brooks et al. (2005) found that climate-related mortality is significantly correlated with 11 key indicators of national vulnerability: (1) population with access to sanitation; (2) literacy rate (15–24-year olds); (3) maternal mortality; (4) literacy rate (over 15 years); (5) calorific intake; (6) voice and accountability; (7) civil liberties; (8) political rights; (9) government effectiveness; (10) literacy ratio (female to male); and (11) life expectancy at birth. These broadly capture the health status of citizens (indicators 1, 3, 5, 11), the quality of education (2, 4, 10) and national governance (6, 7, 8, 9). Other indicators reflecting the state of the economy, infrastructure, technology, agriculture and environment were found to be less important. Ultimately, their choice of metrics was constrained by availability of public domain data from bodies such as the World Bank and United Nations, plus measureable indicators of civil liberties and governance. The results showed that amongst the 59 most vulnerable nations 33 were in sub-Saharan Africa, 5 were small island states, and many had recently experienced conflict.

A decade later many of the same countries (e.g. Burundi, Eritrea and Sudan) still appear as 'hotspots' or near the top of vulnerability tables such as the Notre Dame Global Adaptation Index (ND-GAIN) (Figure 7.3). ND-GAIN is a tiered, composite measure of a country's vulnerability and readiness to adapt to climate change. First tier elements describing vulnerability are ecosystem services, food, health, human habitat, infrastructure, water, adaptive capacity, exposure and sensitivity. Second tier indices include 'dependency on imported energy' (under infrastructure) or 'access to reliable drinking water' (under water). Second tier indices of readiness reflect factors such as 'business regulation and enforcement' (under first tier economic), 'rule of law' (under governance) and 'tertiary education attendance' (under social). Overall, there is a very strong correlation between the level of economic development and climate vulnerability – the most vulnerable and least prepared states are all developing countries (see the top left quadrants in Figure 7.4).

FIGURE 7.3 ND-GAIN index of climate vulnerability in 2014. This is based on factors such as food and water security, health care, ecosystem services, human living conditions, coastal and energy infrastructure. *Source*: ND-GAIN http://index.gain.org/

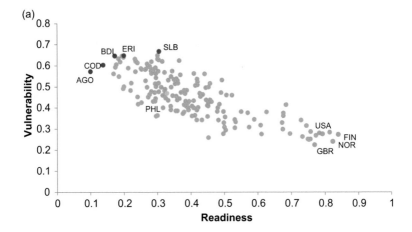

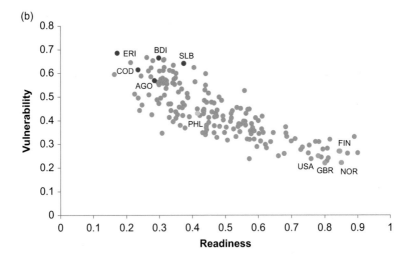

FIGURE 7.4 The ND-GAIN index in (a) 1995 and (b) 2014 for selected countries: Angola (AGO), Burundi (BDI), Democratic Republic of the Congo (COD), Eritrea (ERI), Solomon Islands (SLB); Philippines (PHL); Finland (FIN), Great Britain (GBR), Norway (NOR) and United States (USA). Readiness reflects each country's ability to leverage investment and convert it into adaptation actions. Readiness is measured using three sets of indicators: ability of the business environment to accept economic investment; standard of governance including rule of law, political stability and control of corruption; and social factors such as levels of inequality, education and innovation. *Data source*: ND-GAIN http://index.gain.org/

Changing National Vulnerability

The ND-GAIN index is helpful for tracking changes in national vulnerability and readiness over the period of data availability (1995 onwards). The readiness of most countries has improved over the past two decades (as evidenced by a drift to the right between the upper and lower panels of Figure 7.4). However, this has not been the case for Eritrea, the second most vulnerable and second least ready country in 2014 (Somalia was the most vulnerable). Since 1995 there have been modest improvements in food and water security, and infrastructure capabilities (Figure 7.5); at the same time, Eritrea's readiness to adapt is hampered by governance issues including perceptions about rising corruption, weak regulation of business, and levels of law enforcement/security. On the other hand, the country has become more able to accept investments for adaptation, and has better educational attendance, internet and phone availability (not shown). In comparison, the Philippines were placed mid-table at 95/182 in terms of overall climate vulnerability in 2014 (Figure 7.4). The rank for readiness was 110/184 but is rising due to improving conditions for business, better education and wider availability/use of telecommunications (Figure 7.6). Governance indicators suggest some progress since 2010.

Global Dimensions to Vulnerability

One disadvantage of the ND-GAIN vulnerability and readiness indices is that they only consider stressors and responses operating within the boundaries of a nation state. What is missing is the broader international context; the reality is that globalisation now impacts on local economies, household incomes and hence capacity to adapt. For example, O'Brien et al. (2004) showed how the agricultural sector in India is exposed to climate change along with the negative effects of world price fluctuations and import competition (e.g. due to liberalised trade in edible oils and oilseeds). They produced a globalisation vulnerability map based on a composite of import sensitivity (derived from distance to ports, cropping patterns and crop productivity) and adaptive capacity. This revealed regions of high vulnerability to climate change *with* globalisation in Rajasthan, Gujarat, Madhya Pradesh, southern Bihar and western Maharashtra. The authors of the study suggested that these locations should be focal points for policy intervention.

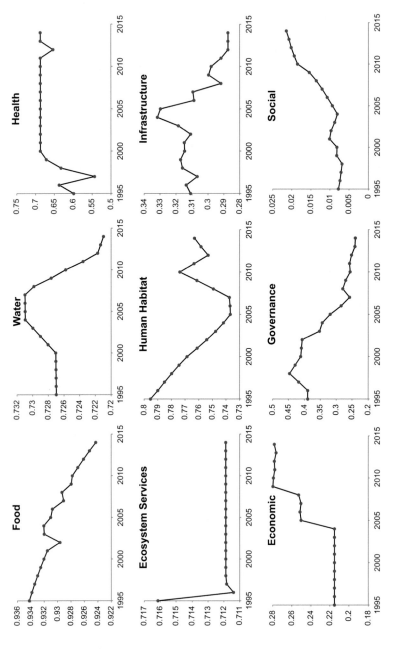

FIGURE 7.5 Indicators of Eritrea's changing climate vulnerability and readiness for adaptation over the period 1995–2014. An improving situation is denoted by lower vulnerability scores (falling trend) and higher readiness scores (rising trend). *Data source:* ND-GAIN http:// index.gain.org/

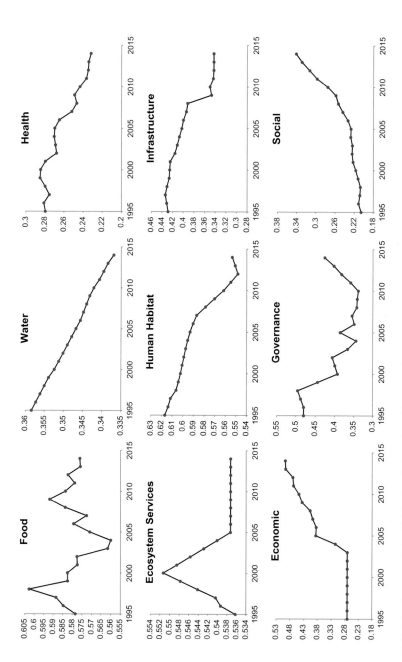

FIGURE 7.6 Indicators of the Philippines' changing climate vulnerability and readiness for adaptation over the period 1995–2014. An improving situation is denoted by lower vulnerability scores (falling trend) and higher readiness scores (rising trend). *Data source:* ND-GAIN http://index.gain.org/

Adaptation Finance

The above example shows how vulnerability indices might shape investment planning performed by development, humanitarian and aid agencies. Indices are also powerful tools for raising awareness of shared vulnerabilities that may, in turn, lead to collective action. For instance, the United Nations Development Programme (UNDP) Climate Vulnerable Forum (CVF) is a diplomatic platform for 20 of the most vulnerable countries (Afghanistan, Bangladesh, Barbados, Bhutan, Costa Rica, Ethiopia, Ghana, Kenya, Kiribati, Madagascar, Maldives, Nepal, Philippines, Rwanda, Saint Lucia, Tanzania, Timor-Leste, Tuvalu, Vanuatu and Vietnam). Under the chair of the Philippines, the CVF intends to negotiate new approaches to climate finance for adaptation and disaster risk reduction. Through the CVF Support Unit there are now opportunities to bring these efforts alongside the strategic objectives for Disaster Risk Reduction envisaged by the Hyogo Framework for Action. This aims to support countries through a raft of measures, including establishment of early warning systems and contingency planning for disasters.

Poverty as a Driver of Climate Vulnerability

This chapter referred to the catastrophic human cost of tropical cyclones to explain why vulnerability to extreme events varies between people and places. The impacts of other climate hazards such as drought, heatwaves and fluvial flooding are also socially and geographically differentiated. Much attention has focused on the fragility of agricultural and coastal livelihoods, but there is growing appreciation of the immense vulnerability of some low-lying megacities. Early warning systems, physical barriers and better spatial planning can go some way to protecting vulnerable people. However, the capacity of households to adapt to future climate variability and change also depends on the strength of social networks and governance systems. Above all, poverty will continue to be a key driver of climate vulnerability and an obstacle to adaptation, whether at the household or national scale. For climate vulnerable countries, the challenge is in implementing strategies that yield sustainable economic development *despite* recurrent climate shocks.

7.2 GROUP EXERCISE: PASTORALISTS AND LONDONERS

NOTES

The aim of this activity is to develop climate vulnerability indices for two contrasting social and geographical contexts. Opiyo et al. (2014) and Wolf and McGregor (2013) should be read before the session. The group is then split in half and sub-divided into sets of two to four students. One half of the group will devise a climate vulnerability index for nomadic pastoralists in East Africa (Figure 7.7, left), the others an index for residents of Central London over the age of 65 (Figure 7.7, right).

Students should begin by identifying the key climate hazards faced by the target groups. Using the template they will then list at least three measureable indicators of climate exposure, sensitivity and capacity to adapt. The sub-components of the livelihood vulnerability index (Table 7.2) are a good starting point for discussions.

After 10–15 minutes collate the suggestions of each group into a master set of climate vulnerability indicators, first for the pastoralists and then for the city-dwellers. These ideas should be compiled by the tutor on a whiteboard or overhead projector. The plenary group will use this information to discuss (i) commonalities between the two sets of indicators; (ii) sources of information, methods and feasibility of data collection; and (iii) ways in which the vulnerability index could be used to inform policy and climate change investments. By the end of the exercise the group should have a deeper understanding of the very different contextual factors that shape climate vulnerability as well as insight to some of the practicalities of their measurement.

FIGURE 7.7 The dwelling of a nomadic pastoralist in Djibouti, East Africa (left) and a view from the London Eye, UK (right).

Worksheet for Pastoralists and Londoners

Indicators	Pastoralists, Djibouti	Older residents, London
Hazards		
Exposure		
Sensitivity		
Adaptive capacity		
Notes		

LEARNING OUTCOMES

- To understand that climate vulnerability is highly contextual because it depends on the level of exposure, sensitivity and adaptive capacity of the chosen receptor(s) to hazards – all of which vary in space and time.
- To recognise that local to global patterns of climate vulnerability are strongly determined by variations in household livelihood strategies and control over assets.
- To appreciate that vulnerability indices are useful tools for first identifying 'hotspots' of susceptibility to climate hazards and then focusing interventions.

7.3 FURTHER QUESTIONS TO RESEARCH AND DISCUSS

Explain the difference between exposure and sensitivity to climate hazards.

What are the strengths and weaknesses of national vulnerability indices such as ND-GAIN?

To what extent could poverty reduction translate into lower vulnerability to climate variability and change?

Evaluate the household as a fundamental unit for assessing climate vulnerability.

7.4 FURTHER READING

Das, S. and Vincent, J.R. 2009. Mangroves protected villages and reduced death toll during Indian super cyclone. *Proceedings of the National Academy of Sciences of the United States of America*, **106**, 7357–7360.

Knutson, T.R., McBride, J.L., Chan, J., et al. 2010. Tropical cyclones and climate change. *Nature Geoscience*, 3, 157–163.

O'Hare, G. 2001. Hurricane 07B in the Godavari delta, Andhra Pradesh, India: vulnerability, mitigation and the spatial impact. *The Geographical Journal*, 467, 234–246.

7.5 OTHER RESOURCES

Climate Vulnerability Monitor http://daraint.org/climate-vulnerability-monitor/climate-vulnerability-monitor-2012/monitor/ [accessed 13/07/16]

Climate Vulnerable Forum (V20) www.thecvf.org/ [accessed 13/07/16]

EM-DAT International Disaster Database www.emdat.be/ [accessed 13/07/16]

Germanwatch Climate Risk Index https://germanwatch.org/en/cri [accessed 13/07/16]

Human Development Index http://hdr.undp.org/en/content/human-development-index-hdi [accessed 13/07/16]

Hyogo Framework for Action www.unisdr.org/we/coordinate/hfa [accessed 13/07/16]

Notre Dame Global Adaptation Index (ND-GAIN) http://index.gain.org/ranking/vulnerability [accessed 13/07/16]

US Naval Oceanography Portal Annual Tropical Cyclone Reports www.usno.navy.mil/JTWC/annual-tropical-cyclone-reports [accessed 13/07/16]

World Risk Report www.worldriskreport.org/ [accessed 13/07/16]

8

When Are Climate Forecasts Good Enough to Take Action?

TOPIC SUMMARY

Seasonal climate forecasts occupy the gap between an extreme weather outlook covering the next few hours to days and a climate model projection for coming decades. They typically give the likelihood of anomalous air temperatures or precipitation 3 to 6 months ahead based on ocean temperatures. This degree of foresight may be long enough to take actions to protect livelihoods and avoid costly emergency responses. However, the humanitarian crisis of the East African drought in 2011 showed that governments, donors and NGOs were reluctant to act on uncertain warnings of extreme weather before total failure of rains became a certainty. Subsequent reviews of the slow international response called for a risk-based approach that reacts to early warnings in staged and proportionate ways. There have also been pleas for more agile humanitarian funding mechanisms to enable early response (rather than manage the aftermath of a crisis). It is acknowledged that the ability to act on a forecast depends as much on the robustness of the underpinning science/technology as the attentiveness of socio-political response systems.

BACKGROUND READING

Dai (2011) provides a comprehensive review of emerging drought risks under climate change. Tisdal (2012) gives a sombre account

cont.

of the unfolding 2011 East African famine. Washington et al. (2006) make a case for deploying seasonal climate forecasting at an early stage in adapting to climate variability and change.

8.1 THERE IS NO SUCH THING AS A CERTAIN CLIMATE OUTLOOK
Connecting Distant Climates

The widely acclaimed book *Late Victorian Holocausts: El Niño Famines and the Making of the Third World* explored links between climate variability and starvation across the British Empire in the late nineteenth and early twentieth centuries. Davis (2001) asserts that the political economy of Liberal Capitalism increased rural poverty and vulnerability to drought. On the other hand, deducing the connection between El Niño and crop failures was only possible because of simultaneous meteorological observations made in colonies separated by thousands of kilometres. The pioneering work of Richard Grove, William Roxburgh, Henry Blanford and Gilbert Walker subsequently established the synchrony of regional droughts linked to an inter-hemisphere atmospheric 'see-saw'. Based on these early insights, it is now well known that when ocean temperatures in the equatorial Pacific are warmer than average (i.e. an El Niño episode) there is heightened risk of drought in India, northeast Brazil, China, western Canada, Ecuador and northern Peru.

In January 2015 weather patterns and sea surface temperatures (SSTs) in the Pacific were indicative of El Niño conditions. A rainfall forecast issued by the International Research Institute (IRI) in early 2015 suggested 45% likelihood of below normal precipitation in northeast Brazil (Figure 8.1, left). [The dry outlook was subsequently corroborated by observations (Figure 8.1, right) although the severity of the actual drought was underestimated by the seasonal forecast.] Given this information, what steps (if any) could have been taken to manage possible socio-economic impacts of a drought by government officials, aid workers, community leaders or farmers in the affected area? The dilemma faced by all such decision-makers is that the underpinning science of the forecast systems is imperfect and, even if the outlook likelihood (45%) was completely trustworthy, the threshold for action depends on attitudes to risk and on availability of resources to respond.

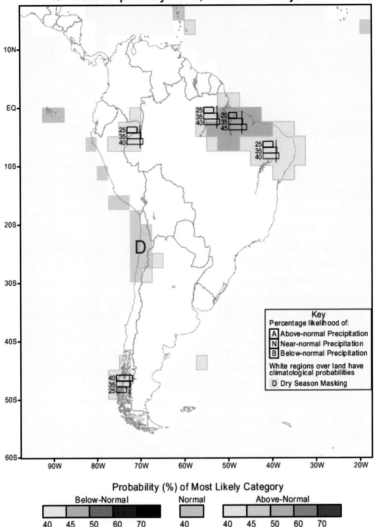

(a) **IRI Multi-Model Probability Forecast for Precipitation for March-April-May 2015, Issued January 2015**

Key
Percentage likelihood of:
A Above-normal Precipitation
N Near-normal Precipitation
B Below-normal Precipitation
White regions over land have climatological probabilities
D Dry Season Masking

Probability (%) of Most Likely Category

Below-Normal — 40 45 50 60 70
Normal — 40
Above-Normal — 40 45 50 60 70

FIGURE 8.1 (a) A seasonal precipitation forecast for South America issued by IRI in January 2015 for the period March to May 2015. Numbers next to bars show the percentage likelihood of above-normal (top value), near-normal (middle value) or below-normal (lower value) precipitation. The long-term average is 33% for each category so a large departure from 33% suggests a strong likelihood of that category of above/near/below-normal precipitation. (b) Actual rainfall during the forecast period March to May 2015 showing overall underestimation of the spatial extent of the region affected by drought. *Source*: Figures courtesy of the International Research Institute for Climate and Society, Columbia University, USA. http:// iri.columbia.edu/our-expertise/climate/forecasts/seasonal-climate-forecasts/

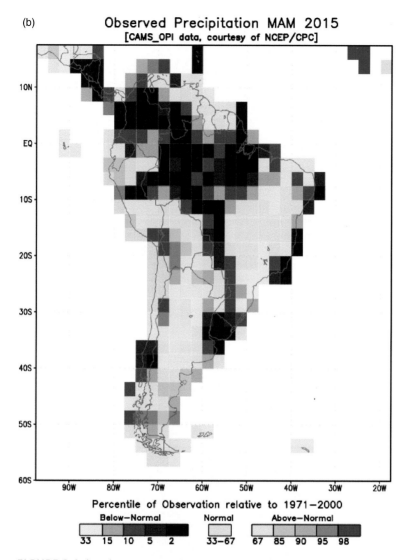

FIGURE 8.1 (*cont.*)

East African 2011 Drought

These tensions were exposed most cruelly during 2011 after a cool-water La Niña event in the Pacific caused severe drought across the Horn of Africa (HOA) (Figure 8.2). Even in March 2010 early warning systems (EWS) were signalling critical levels of acute malnutrition. By the time the United Nations declared famine in July 2011, levels

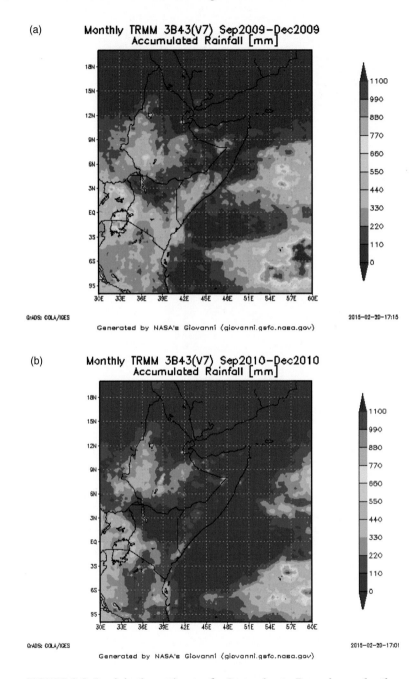

FIGURE 8.2 Precipitation estimates for September to December under the
(a) 2009 El Niño and (b) 2010 La Niña conditions. Note the large difference
over southern Somalia coinciding with the area of catastrophic famine
shown in Figure 8.3. *Data source*: Tropical Rainfall Measuring Mission (TRMM)

of malnutrition and mortality were above catastrophic levels (Kim and Guha-Sapir, 2012). At that point, up to 12 million people were starving and in need of humanitarian assistance (Figure 8.3). Surveys conducted in southern Somalia showed that the majority of locations had malnutrition rates exceeding 30% and crude mortality rates greater than the emergency threshold of 1/10,000 per day (Kim and Guha-Sapir, 2012). Until then, aid efforts had been insufficient to avert the crisis. Following the UN famine declaration there was a surge in humanitarian aid and media interest, but it was too late. By the end of the drought up to 100,000 people had died. Jan Egeland (a former UN Emergency Relief Coordinator 2003–2006) referred to this avoidable loss of life and hardship as an 'outrage' (Save the Children and Oxfam, 2012).

Hard Lessons Learnt

The disaster prompted international governments to draft the *Charter to End Extreme Hunger*.[1] This set out steps to avoid a repeat of the HOA humanitarian crisis including measures to strengthen links between early warning systems and mechanisms for timely and appropriate responses. Some of the proposed actions are procedural, such as better coordinating the appeal timeline with knowledge of the typical timing of critical rains (combined with information from seasonal forecasts). For instance, if a relief assessment is too early it may not consider possible failure of monsoon rains, thereby underplaying the chance of rapidly deteriorating food security. Judgements about when to mobilise resources can be automated and sped up using trigger points for action (Kim and Guha-Sapir, 2012). This requires indices and thresholds of need that reflect actual hardship. For example, Ethiopia has the Productive Safety Net Programme (PSNP), which transfers cash to food-insecure households in return for public works when rainfall is deemed insufficient to meet the water requirements of staple food crops (Bastagli and Harman, 2015).

Seasonal Climate Forecasts

Seasonal climate outlooks complement these hunger-based surveillance systems by forewarning of the approach of critical rainfall

[1] https://hungercharter.wordpress.com/charter-to-end-extreme-hunger/

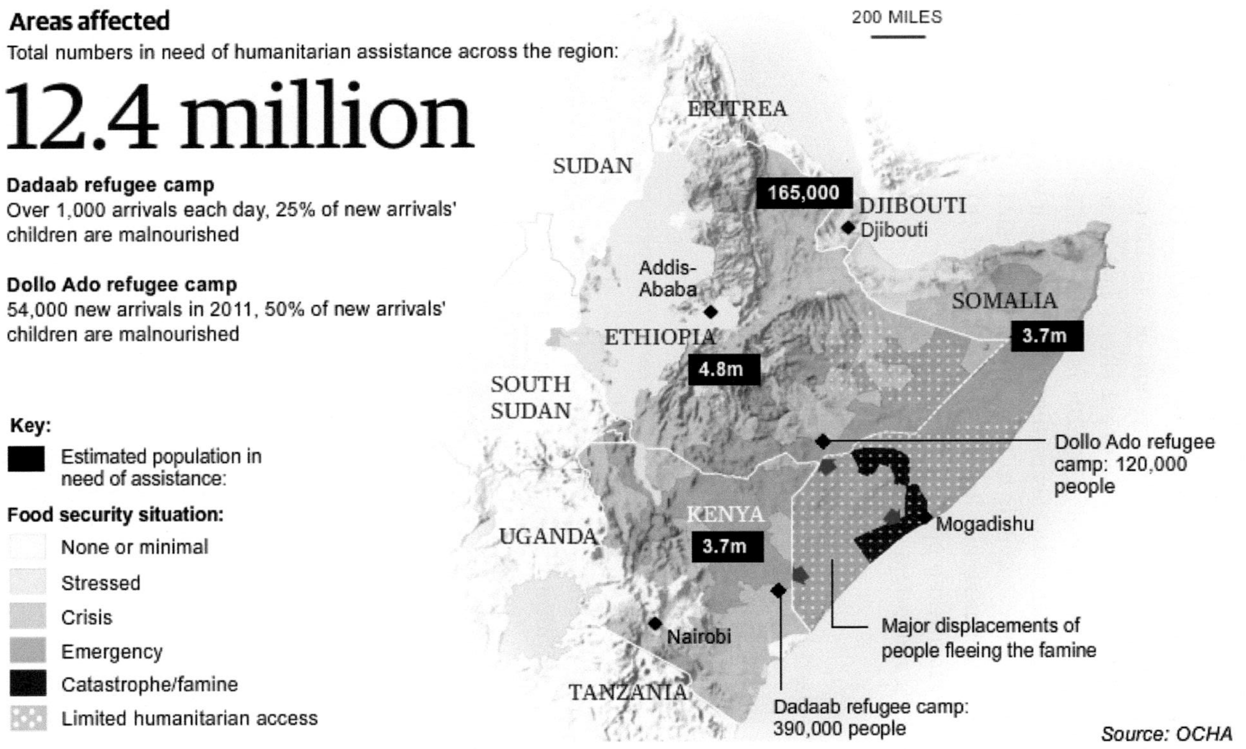

FIGURE 8.3 Consequences of drought in East Africa in 2011. *Source:* United Nations Office for the Coordination of Humanitarian Affairs (OCHA). Copyright Guardian News and Media Ltd 2016.

thresholds. Forecasts may be issued 3 to 6 months ahead based on comparatively simple statistical relationships (e.g. Nicholson, 2014) or complex dynamical models (akin to global climate models; see Dutra et al., 2013) that link SSTs to regional rainfall and temperature anomalies. Both techniques rely on the fact that slowly evolving ocean temperatures 'imprint' themselves on more rapidly changing weather patterns. For example, there is a weak negative correlation between the boreal autumn 'short rains' in Djibouti and SSTs in the Pacific (Figure 8.4a). This means that when cool waters (i.e. La Niña conditions) are observed (as in Figure 8.4b) there is increased likelihood of drought in Djibouti and the HOA more generally. Conversely, during El Niño years there are heavier rains, more frequent flood episodes and related outbreaks of waterborne disease in the HOA.

Evaluation of the European Centre for Medium Range Weather Forecasts (ECMWF) for the 2010–2011 drought in the HOA

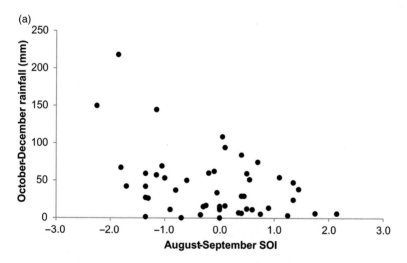

FIGURE 8.4 (a) Empirical relationship between August–September Southern Oscillation Index (SOI) and following October–December rainfall measured at Djibouti airport 1951–2000. Negative SOI denotes El Niño conditions; positive SOI occurs under La Niña conditions. Meteorological observations show that cool SSTs in the Pacific (La Niña) are typically associated with less autumn/winter rainfall in Djibouti. (b) The pattern of Pacific SST anomalies signalling cool La Niña conditions in November 2010. *Source*: NOAA Operational SST Anomaly Chart for 11 November 2010, www.ospo.noaa.gov/data/sst/anomaly/2010/anomnight.11.11.2010.gif

NOAA/NESDIS 50 KM GLOBAL ANALYSIS: SST Anomaly (degrees C), 11/11/2010
(white regions indicate sea-ice)

(b)

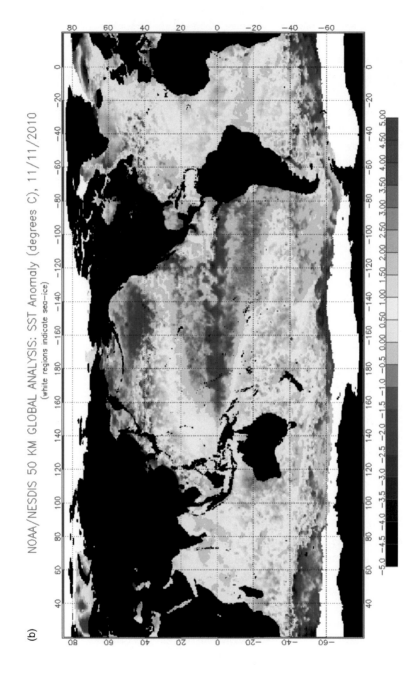

FIGURE 8.4 (cont.)

revealed that the development of a La Niña event from June 2010 onwards was predictable (Dutra et al., 2013). Although drought was expected from July 2010 to year end, ECMWF seasonal outlooks did not foresee continuation of the anomaly into March–May 2011 (except for forecasts issued in March 2011). This is because March–May rains in the HOA are not strongly related to preceding SSTs in the Pacific. Likewise, Nicholson (2014) found the same 'predictability barrier' for spring and summer 'long rains' over the HOA using a statistical model based on atmospheric (rather than SST) variables. Others assert that predictability can be improved for the long rains using more elaborate indices of SSTs in the western Pacific and Indian Ocean (Funk et al., 2014). Even then, reported forecast skill is still no better than 40% of observed rainfall variability.

Drought Monitoring and Decision-Making

Prototype drought monitoring and forecasting systems are being developed in ways that improve the relevancy of information to users in agriculture, water and health sectors (e.g. Kumar et al., 2004; Mwangi et al., 2014; Shukla et al., 2014). For instance, dry and dusty conditions are known to influence the timing and location of meningitis epidemics in sub-Saharan Africa (Sultan et al., 2005) – a fact that is exploited by the African Center of Meteorological Application for Development (ACMAD) meningitis bulletin.

Likewise, the African Drought Monitor (ADM) provides real-time information and advance warning based on a blend of remotely sensed water balance, crop and vegetation indices, hydrological modelling, downscaling techniques and seasonal forecasts (Sheffield et al., 2014). This system produced skilful forecasts of declining soil moisture across East Africa in the last 6 months of 2010 (Figure 8.5a) but, like ECMWF, failed to show persistence of the drought towards its climax in April–June 2011 (Figure 8.5b). More generally, Regional Climate Outlook Forums (RCOFs) were established by the World Meteorological Organisation (WMO) in the HOA and Southern Africa to collate seasonal prediction information and deliver expert-based, consensus outlooks in user-relevant formats.

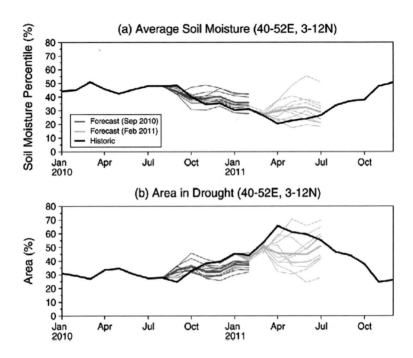

FIGURE 8.5 Six-month hydrological forecasts issued during the 2010/11 Horn of Africa drought. (a) Soil moisture percentile averaged over the main drought region (3°–12° N, 40°–52° E) from the historical hydrological model (VIC) simulation, September 2010 forecast (red lines), and February 2011 forecast (green lines). (b) Percentage area in drought based on a drought threshold of the 20th percentile of soil moisture for the historical simulation and the two forecasts. Forecasts are shown for 10 ensemble members and for the ensemble mean (thick lines). *Source*: Sheffield et al. (2014) ©American Meteorological Society. Used with permission.

Acting on Uncertain Forecasts

Given these resources and early warning of an impending crisis, why was the international response to the HOA famine so inadequate? Several explanations were offered by Save the Children and Oxfam (2012). First, there was fear of financial and reputational damage of getting it wrong, or of being too interventionist, or of undermining communities' ability to cope with recurrent drought episodes. Second, media and public attention were not engaged until the situation had reached crisis point. Third, there was reluctance on the part of affected governments to declare an emergency for fear of appearing weak. Finally, knowledge of local conditions on the ground was not communicated in an effective and timely way to managers responsible for releasing drought-relief funds.

The *Charter to End Extreme Hunger* founded in 2011 presses for a fundamental shift in attitude from a reactive to proactive approach to dealing with climate hazards. This means acting despite uncertainty in forecasts by managing risk rather than a crisis. In this way, the principles of risk reduction would lead to a staged response to an unfolding drought (Save the Children and Oxfam, 2012). Early warnings might trigger advocacy and fund raising. As the outlook worsens, livelihood protection measures could be invoked (such as supplementary feed and vaccinations for livestock, rehabilitation of local water schemes, cash/food for work programmes). Finally, a food/nutrition response would be implemented. However, early interventions are actually more cost-effective than disaster relief. For example, restocking sheep and goats in the Afar region of Ethiopia after a drought is six times more expensive than keeping herds alive by supplementary feeding (see also Aklilu and Wekesa, 2002). Yet this level of integration requires strong partnerships between climate information users and suppliers and effective two-way communication (Hellmuth et al., 2011). Long-term development programmes and humanitarian strategies also need to be aligned in ways that build overall resilience to climate shocks.

Adapting better to *present* climate variability is regarded by some as the first step towards addressing the greater development and scientific challenges posed by climate change (Washington et al., 2006). Seasonal climate forecasting is one decision-support tool that could improve (i) preparedness in agriculture (for timing of planting, fertiliser and pesticide application, or irrigation scheduling); (ii) resource management (e.g. wildfire, rangelands, reservoir storage and hydropower); and (iii) contingency planning (for extreme events, disease control and major civil construction projects). However, this involves acting on uncertain information from early warning systems with lead times of weeks to months ahead. Even if the skill of early warning technologies improves, the value of this information can only be fully realised when accompanied by flexible humanitarian funding mechanisms and responsive organisational structures.

Embracing Uncertainty

No forecast (whether for the next few days, seasons or decades ahead) is ever 100% certain. Therefore, difficult judgements have to be made about whether to act on this imperfect information or to wait and see. Although the climate outlook is uncertain at all time scales, there are still plenty of sensible measures that can be taken to reduce vulnerability ahead of harmful weather events. Knowing who or what is most vulnerable to climate is a useful starting point (Chapter 7). Assessing the changing patterns of climate risk at national (Chapter 9) to sub-regional or even city scales (Chapter 10) is an early step towards formulating an adaptation plan (Chapter 14). Predicting then acting is an appealing formula, but whether for climate projections, downscaled scenarios or seasonal forecasts, uncertainty is a fact of life. Therefore, these powerful tools must always be used alongside other instruments for managing climate risk, such as spatial planning, improved resource efficiency and institutional capacity building.

8.2 GROUP EXERCISE: *PAYING FOR PREDICTIONS*

NOTES

This activity is a variant of the game *Paying for Predictions* conceived by Suarez and Mendler de Suarez (2012). Their games-based approach to experiencing climate risk has even been taken to the White House.

The purpose of *Paying for Predictions* is to give players an insight to the costs and benefits of using (or ignoring) seasonal climate forecasts. The facilitator assumes the role of the 'forecast provider' whereas players emulate the decisions taken by humanitarian agencies about whether to be proactive or reactive to an impending flood or drought. There is always a cost associated with taking anticipatory action (compared with doing nothing). For example, hardier varieties and fertilisers might be purchased before a drought to produce a crop, whereas emergency shelter and supplies could be distributed ahead of an expected flood. There may then be significant payback if cropping regimes or cultivars are better matched to drought conditions or a humanitarian crisis is avoided in the event of a flood.

Teams should be formed of four to six players. Each team has one red dice and one white dice and the worksheet below. In the first two rounds teams do not have access to seasonal forecasts; decisions must be made on guesswork/probabilities alone. From the third round onwards, teams can purchase seasonal forecasts which give them foresight of the regional climate outlook (the red

cont.

dice) but not of local conditions (the white dice). Based on this additional information teams have improved odds of taking correct action before the outcome is known. Play proceeds as follows:

Practice rounds: The regional climate is represented by a red six-sided dice; the local climate by a white six-sided dice. Before any dice are rolled each team must declare whether they will be proactive (pay 1 token if expecting flood; 2 if expecting drought) or hope for the best (pay nothing). Both dice are then rolled: drought is indicated by a combined score of 4 or less; flood if the total is 10 or more; average conditions by 5 to 9. If a drought is indicated and a team was proactive they do not have to pay for humanitarian relief; if they made no preparations they pay the cost of famine assistance, which is 4. If a flood is indicated and the team was proactive they pay no additional costs. If no preparations were made the costs of flood relief are 3 tokens. If average weather, there are no costs or benefits from the event, but unnecessary flood and drought preparations still cost the team 1 and 2 tokens respectively. Teams record dice scores and the cost–benefit outcome on their worksheet.

Play rounds: Teams now have the extra option of obtaining a seasonal forecast for 1 token. If the forecast is purchased the team may roll the red dice *before* they declare their disaster risk reduction strategy (i.e. do nothing, prepare for drought or prepare for flood). Teams that decide not to buy the forecast proceed as in the practice round. Clearly, purchasing the forecast may be advantageous in the event that a high probability of extreme weather is indicated (e.g. if the red dice shows 1, players know that they have a 3 in 6 chance of drought, and nil chance of flood). On the other hand if the forecast is near average then the outlook is less determinate (e.g. if the red dice shows 3 there is still a 1 in 6 chance of drought, or if red is 4 then there is a 1 in 6 chance of flood).

Game over: Time permitting, up to 10 rounds should be played with the seasonal forecast capability. Final scores are calculated by summing the final column of the worksheet and the winning team has the lowest accumulated costs.

Game variants: In a simpler version of the game, only the facilitator rolls the dice with team representatives coming forward to purchase the forecast (i.e. to view the red dice before the white is thrown). Teams should declare their intended action using a

cont.

suitable symbol such as a red (drought), green (no action) or blue (flood) card. In a more complex version, climate change is introduced (secretly) mid-way through the game by replacing the regional dice with an eight-sided dice. This increases the likelihood of flood events throughout the rest of the game.

The designers of the original game suggest that the facilitator keeps teams time-pressed by counting down (10, 9, 8, … STOP) to each decision (i.e. purchase forecast, declare plan). This simulates the narrow window of opportunity for reviewing then acting (or not) on a seasonal forecast; as well as the limited time and resources available to humanitarian agencies and governments to procure and interpret such products.

At the end of the game teams should reflect on the effectiveness of their strategy and report back any insights gained on using seasonal forecasts in practice.

Paying for Predictions*

Team name:
Team members:

Worksheet								
Round	Buy forecast?		Declare plan (tick one)			Outcome (dice roll)	No Action Cost	Cost
	Yes Pay 1	No Pay 0	Drought Pay 2	No action Pay 0	Flood Pay 1	Drought (≤4) Average (5-9) Flood (≥10)	Drought (4) Average (0) Flood (3)	
Example			2			7 (average)	0	2
Example	1				1	10 (flood)	0	2
Practice								
Practice								

cont.

* Adapted from Suarez and Mendler de Suarez (2012)

	Buy forecast?		Declare plan (tick one)			Outcome (dice roll)	No Action Cost	
Round								Cost
	Yes Pay 1	No Pay 0	Drought Pay 2	No action Pay 0	Flood Pay 1	Drought (≤4) Average (5-9) Flood (≥10)	Drought (4) Average (0) Flood (3)	
1								
2								
3								
4								
5								
6								
7								
8								
9								
10								
Total								

(cont.)

Questions to Discuss

What was the team's overall strategy?

What insights were gained about the utility of seasonal forecasts?

LEARNING OUTCOMES

- To appreciate that seasonal climate forecasts are uncertain, with predictive skill depending on region, season, variable (generally less for precipitation than temperature) and forecast horizon (lower confidence for longer leads).
- To understand that reticence to act on a forecast can be due to a mix of political and institutional obstacles as well as the paralysing effect of uncertainty.
- To recognise that early warning systems are part of a larger 'tool kit' of technologies and strategies that can strengthen societal resilience to climate variability and change.

8.3 FURTHER QUESTIONS TO RESEARCH AND DISCUSS

Explain how early warning systems improve overall resilience to climate variability now and in the 2050s?

What are the advantages of managing the risk rather than the consequences of a climate-related humanitarian crisis?

What institutional obstacles might impede the use of seasonal forecasts for humanitarian assistance?

8.4 FURTHER READING

Maskey, S., Toth, E., Opere, A., et al. (eds.) 2015. Drought forecasting and warning. *Hydrology and Earth System Sciences*, Special Issue. www.hydrol-earth-syst-sci.net/special_issue186.html.

Mendler de Suarez, J., Suarez, P. and Bachofen, C. (eds.) 2012. *Games for a New Climate: Experiencing the Complexity of Future Risks.* Frederick S. Pardee Center for the Study of the Longer-Range Future, Boston University.

8.5 OTHER RESOURCES

African Center of Meteorological Application for Development (ACMAD) www.acmad.net/new/ [accessed 13/07/16]

International Research Institute for Climate and Society Seasonal Climate Forecasts http://iri.columbia.edu/our-expertise/climate/forecasts/seasonal-climate-forecasts/ [accessed 13/07/16]

Met Office decadal climate forecast www.metoffice.gov.uk/research/climate/seasonal-to-decadal/long-range/forecasts [accessed 13/07/16]

NOAA Operational SST Anomaly Charts www.ospo.noaa.gov/Products/ocean/sst/anomaly/ [accessed 13/07/16]

Red Cross/Red Crescent Climate Centre www.climatecentre.org/ [accessed 13/07/16]

Regional Climate Outlook Forums http://public.wmo.int/en/our-mandate/climate/regional-climate-outlook-products [accessed 13/07/16]

Who or What Is Most at Risk from Climate Change?

TOPIC SUMMARY

Risk is defined as the likelihood of an event multiplied by the consequence(s) of that event. Hence, risk assessment is a powerful tool for anticipating threats and opportunities presented by climate change. This informs a proactive rather than reactive approach to future events. The assessment can be applied to a range of impact receptors such as people, places, species, or parts of the natural and built environment. These may be grouped into units of assessment covering households, biomes or sectors. The scale of analysis may be individual sites, cities, nations, or even global. Five ingredients are needed to undertake a climate risk assessment: (i) frameworks to guide the process; (ii) clearly defined receptors and units of interest; (iii) physical and social scenarios of change; (iv) tools for translating climate hazards into consequences for the receptors/units; and (v) indicators of climate impact that are relevant to decision-makers. The goal is to highlight risks of particular concern and thereby shape a policy or plan of action. This cannot be accomplished by treating climate threats in isolation of other non-climatic drivers of change in the receptor.

BACKGROUND READING

Scholze et al. (2006) provide a global-scale risk analysis of climate driven changes in key ecosystem processes during the twenty-first century to identify hotspots of biome change due to wildfire and

cont.

freshwater runoff variations. At the opposite end of the spectrum, Brekke et al. (2009) undertake a local-scale risk assessment for reservoir operations in California's Central Valley Project and State Water Project systems.

9.1 ANTICIPATING FUTURE THREATS AND OPPORTUNITIES
Birds and the Trees

No one is entirely sure why the Forêt du Day is dying. Remnants of a once extensive dry tropical forest are hidden amongst the deep ravines of the Goda Mountains in Djibouti (Figure 9.1). Estimates suggest that 88% of the forest has been lost in the last 200 years (Comité Nationale pour l'Environnement, 1991) with 95% of the lingering juniper either dead or dying (Bealey and Rayaleh, 2006). This is bad news for the critically endangered Djibouti francolin (*Pternistis ochropectus*), a pheasant-like bird found only in this forest. A survey in 2009 estimated that there were probably fewer than 600 individuals left, now split into two isolated populations. Explanations for the forest die-back include livestock grazing and exploitation of trees for firewood or timber, but many villagers living in the area blame the demise of the forest on drought beginning in the 1990s (Fisher et al., 2009). Their perceptions match data showing that the seasonal distribution and inter-annual variability of rainfall has changed over large swathes of the dry tropics (Feng et al., 2013).

Case Study: Natural Hazards and Vulnerability of Djibouti

International donors are working alongside the Government of Djibouti to assess the wider threats posed by climate variability and change, in order to strengthen the national capacity to understand

FIGURE 9.1 Juniper stress in the Forêt du Day, Goda Massif, Djibouti.

and adapt to such risks (e.g. Mora et al., 2010; Wilby, 2009). Djibouti is a small, resource-poor nation, strategically located in the Horn of Africa at the southern end of the Red Sea. The country has limited arable land (just 0.1% by area), rainfall and groundwater reserves. According to the United Nations' 2014 estimate, Djibouti's population is ~886,000, of which 59% live in the capital Djibouti-Ville. The hinterland, an extension of the deserts of Ethiopia and Somalia, is sparsely occupied by poor pastoral and nomadic populations. Djibouti also has a large expatriate community of foreign military personnel, technical assistance staff and their families, and private business. About 34% of Djibouti's population are under age 15 and 6% over 60. Life expectancy at birth is 61 years. By comparison, Norway has 19% below age 15, 22% above 60, and life expectancy of 82 years.

Djibouti is exposed to a range of natural hazards, including: (i) multi-year droughts that compound the effects of the natural aridity of the climate and exacerbate water scarcity; (ii) flash floods with attendant loss of life, waterborne diseases, impact on livelihoods and damage to infrastructure, as in 1994; (iii) earthquakes typically in the region of magnitude 4 but with potential for magnitude 7 on the Richter scale, which could liquefy the soils of the Oued Ambouli delta on which Djibouti-Ville sits; (iv) volcanism along the Afar rift that could have direct consequences for local economic activities and indirectly on heavily populated areas; and (v) fires fuelled by the protracted dry periods. There is a strong association between the occurrence of La Niña and regional drought (Figure 8.4a), or between El Niño, heavy rainfall, local flooding and outbreaks of waterborne diseases (Table 9.1).

Vulnerability to hazards is increased by weak governance of water resources, together with environmental degradation (such as soil and water contamination), and where there is a low capacity for land use planning, building codes, social-environmental and financial protection schemes, and public policies for preventive disaster risk management. These issues are matters for concern regardless of the longer-term threats posed by climate change. For example, climate variability and change could reduce groundwater availability, but Djibouti is already facing a water crisis due to the demand placed on limited renewable freshwater resources by population growth. With less than 400 m^3/yr per capita, the country is even now classified as water scarce (according to the World Health Organisation definition of <1000 m^3/yr per capita) (Wilby, 2009).

TABLE 9.1 Natural hazards recorded by the Centre d'Études et Recherches de Djibouti (CERD) (up to November 2008) and their association with El Niño (E), La Niña (L) and Neutral (N) conditions in the Pacific Ocean. Adapted from Wilby (2009).

Risk Assessment and Reduction

Risk is defined here as *"the likelihood over a specified time period of severe alterations in the normal functioning of a community or a society due to hazardous physical events interacting with vulnerable social conditions, leading to widespread adverse human, material, economic, or environmental effects that require immediate emergency response to satisfy critical human needs and that may require external support for recovery"* (IPCC, 2012:5) (Figure 9.2). In short, risk is the combination of the probability of a hazardous event and associated consequences. Since year 2000, the Government of Djibouti has been developing a legal and institutional framework to address the risks posed by natural hazards such as drought. It is recognised that risk reduction and hazard management are integral to development planning and poverty reduction. This involves preparing national strategies for risk and disaster management, including staffing, training, dissemination of warnings, and monitoring equipment. All such activities are compatible with the objectives of reducing vulnerability and increasing resilience to climate variability and change.

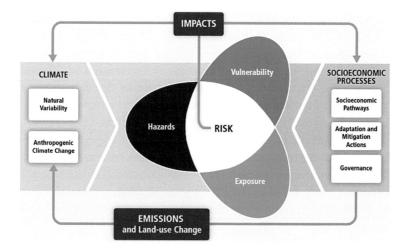

FIGURE 9.2 Risk of climate-related impacts resulting from the interaction of climate-related hazards (including harmful events and trends) with vulnerability and exposure of human and natural systems. Changes in both the climate system (left) and socio-economic processes (right) are drivers of risk. *Source*: IPCC (2013a)

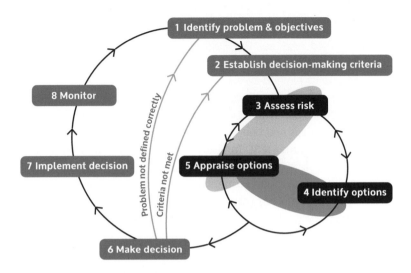

FIGURE 9.3 A framework to support good decision-making in the face of climate change risk. *Source*: Willows and Connell (2003)

A Risk Assessment Framework

Climate risk assessment involves a deliberate and systematic screening of hazards and consequences for risk receptors (e.g. people, species, infrastructure) and units (e.g. places, households, sectors). The UK Climate Impacts Programme (UKCIP) risk and uncertainty decision-making framework (Figure 9.3) has been widely used and is endorsed by the United Nations Framework Convention on Climate Change.[1] The framework guides decision-makers through an eight stage process intended to gauge the significance of climate risks and opportunities relative to non-climatic threats (Willows and Connell, 2003). The overall objective is to identify the most appropriate adaptation measure(s) to achieve intended outcome(s), such as fewer fatalities from flash flooding despite more intense rainfall in the future.

Stages 1 and 2 are about structuring the problem and establishing the markers of success from the outset. Stage 3 is the formal risk assessment based on expected climate scenarios, receptor vulnerabilities and exposure (Figure 9.3). Stages 4 and 5 identify sets of adaptation options that can then be tested, before making then implementing decisions in Stages 6 and 7, respectively. Stage

---

[1] http://unfccc.int/adaptation/nairobi_work_programme/knowledge_resources_
   and_publications/items/5496.php

8 recognises the need to monitor and evaluate the decision in order to adjust to any evolving climate and non-climatic stresses. The whole process is open-ended and iterative with the possibility of revisiting a decision in the light of new information. A tiered approach is also taken to ensure that the time and resource invested in assessment are proportionate to the scale of the risks and decisions that have to be made. The following sections illustrate how Stages 1 to 3 work in practice using the Djibouti climate risk assessment as an example. Stages 4 to 8 will be picked up later under discussions about adaptation in Chapter 14.

Stage 1: Identify Problems and Objectives

The motivations behind a risk assessment help to frame the overall scope and approach adopted in subsequent stages. The primary objective of the Djibouti study was to review climate risks, assess priorities for action, then devise a programme of further work to address key knowledge gaps. Ultimately, the evidence gathered would inform the National Adaptation Programme for Action (NAPA) on Climate Change. The high vulnerability of both rural and urban populations to existing climate hazards has already been noted. However, these threats must be viewed alongside a host of non-climatic pressures such as population and economic growth, environmental degradation and technological change. Hence, a strategy for addressing water scarcity would have to consider both the climatic and non-climatic drivers of the supply–demand balance. The point is that the most climate-sensitive decisions have to be recognised from the outset.

The key stakeholders were directors of government departments responsible for developing long-term plans for priority sectors such as agriculture, health and water. Although the Djibouti NAPA looks forward to the 2050s, short-term benefits are expected during the interim. For example, strengthening early warning systems for flash flooding and storm surge would reduce damages now as well as in the future. Other time scales for implementing decisions have a longer horizon, as is the case for protecting coastal communities and habitats from sea level rise. In many situations there will be multiple objectives and resource constraints, so the decision-maker may trade off actions on far-term for near-term investments. This is likely given the larger uncertainty attached to distant climate change scenarios (Figure 6.4).

Stage 2: Establish Decision-Making Criteria

Having structured the problem and specified the assessment objectives, the next step is to set down the decision criteria. This involves identifying the receptors and exposure units. For example, pastoralists (receptors) suffering the most extreme poverty are found in the southern (Ali-Sabieh) and western (Dikhil) districts (units) of Djibouti. These groups and places provide a legitimate focus for an assessment of drought risks to pasture lands and traditional livelihoods based on extensive livestock herding (Figure 7.7, left). Levels of extreme poverty (defined as US$1.80 per day) amongst these communities already exceed 70% of the population (Direction de L'Aménagement du Territoire et de l'Environnement [DATE], 2006). Therefore, intervention success might be measured by rapid poverty reduction against this baseline by a specified date (or end-point), despite climate variability and change. Many factors can shape the decision about the end-point outcome, including gender, social status, income and age of the target group. In addition, the legislative context and any resource constraints should also be taken into account. In practice, the choice of receptor and risk assessment end-point is often defined by data availability for benchmarking the starting point and tracking progress (Section 7.2).

Stage 3: Assess Climate Risks

By now the investigator has established the terms of reference and intended outcomes of the decision-making process. The main purposes of Stage 3 are to characterise the risk; provide qualitative or quantitative estimates of the risk; assess the consequences of uncertainty for decisions; and compare sources of risk, including climate and non-climatic threats (Willows and Connell, 2003:18). These involve defining the time horizon of the decision and most important climate variables. The choice of tools and techniques for creating the climate scenarios must be matched to the intended application, taking into account local constraints of time, resources, human capacity and supporting infrastructure (Wilby et al., 2009). The options might range from qualitative narratives of plausible climate changes (perhaps based on analogues from the past) through to regional climate downscaling.

Seasonal forecasts of rainfall deficits or soil moisture could be used to assess the risk of famine and thereby trigger humanitarian relief to the most vulnerable groups of pastoralists (Chapter 8). Over coming decades, livelihoods may be threatened by changes in the duration of future dry spells, maximum rainfall intensity, and maximum temperatures (Figure 9.4). These (highly uncertain) climate outlooks could translate into, respectively, drought risk to pastures and watering holes, flash flood risk to people and property, or human and livestock health. Quantitative risk assessments of these

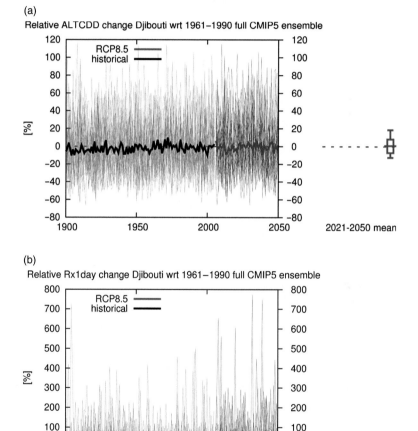

FIGURE 9.4 Changes in annual (a) maximum dry-spell length (%), (b) maximum 1-day precipitation total (%) and (c) maximum daily maximum temperature (°C) projected by the CMIP5 ensemble for Djibouti under business as usual emissions (RCP8.5) relative to the 1961–1990 baseline.

(c)

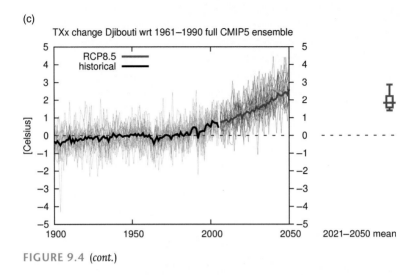

FIGURE 9.4 *(cont.)*

consequences are only possible where there are data on attendant impacts to establish cause and effect (as in Table 9.1).

Even where such data exist, interpretation may be confounded by other non-climatic factors. For example, households with more diverse income streams may be buffered against climate shocks to pastoralism. Conversely, families caring for members with chronic health issues may be more sensitive to heat and drought. As expected, there tends to be less data for some of the most vulnerable people and places. Therefore, the risk assessment may revert to a desk-top review of literature or expert judgement of the estimated likelihood and severity of impacts. The case study shows how a mixed quantitative and qualitative approach was adopted in a multi-sectoral climate risk assessment for Djibouti (Mora et al., 2010; Wilby et al., 2010a).

Case Study: Climate Risks Faced by Djibouti

According to the NAPA, the most vulnerable sectors in Djibouti are water resources, agriculture and forestry, livestock, coastal zone and marine ecosystems (DATE, 2006). Table 9.2 provides a summary of climate change drivers and associated impacts, by sector. In many cases, climate change is expected to amplify existing risks. For example, extraction of riverine sand and clay for construction increases rates of flow over exposed bedrock channels, and with this come more downstream erosion and flash flooding. Coral reefs that

TABLE 9.2 Potential climate change impacts on Djibouti. *Source*: Wilby (2009)

Sectors	Drivers	Example impacts
Water resources	Climate variability, flooding and drought	Lack of potable water; rising salt concentrations; falling water tables; contamination of wells by floodwaters carrying suspended materials; increased grazing pressure around watering holes; reduced groundwater recharge during spates
Agriculture, forestry and livestock	Climate variability, flooding and drought	Loss of yields due to water scarcity or destruction by flooding; disease or pest outbreaks; loss of soil fertility due to salinisation; soil erosion by wind and runoff; reduced area of pasture; increased mortality and morbidity of livestock; abandonment of terraces; increased reliance on emergency food aid and assistance
Human health	Climate variability, flooding and drought, rising temperatures	Loss of life due to flash floods; outbreaks of diarrhoea, cholera and malaria; heat stress; reduced food security
Built environment	Flooding	Damage to property and national infrastructure; increased surface and groundwater contamination; disruption of surface transportation networks

cont.

TABLE 9.2 (cont.)		
Sectors	Drivers	Example impacts
Terrestrial ecosystems	Climate variability, flooding and drought	Loss or degradation of natural habitats; replacement of endemic tree and herbaceous species by invasives such as *Prosopis (Mesquite)*; loss of vegetation cover increases runoff and reduces recharge
Coastal zone	Sea level rise	Loss of coastal land due to inundation and erosion by the sea; loss of agricultural output; loss of mangroves and fish spawning habitat
Marine ecosystems	Rising ocean temperatures	Increased coral bleaching and disease; appearance of alien species; changing fish stocks in shallow waters

are already impacted by pollution, over-fishing and tourism may be further stressed by rising ocean temperatures. Aquifers depleted by over-exploitation may recharge less effectively during more intense rainfall–runoff events in the future.

As noted before, even without climate change, Djibouti is facing a water resource crisis. Per capita water availability is estimated to be <400 m^3/yr; well below the average for the Middle East and North Africa (1250 m^3/yr). If the population reaches 1 million, and the renewable resource remains unchanged, per capita water availability would fall to 300 m^3/yr. Available supplies are placed under further stress by groundwater contamination from untreated wastewater, by accidental spills, and saline intrusion to coastal aquifers. Since the 1960s the concentration of chloride has trebled in Djibouti's main aquifer, a tell-tale sign of encroaching salt water (Houssein and Jalludin, 1995).

There have been only a handful of studies looking specifically at climate change impacts on Djibouti (e.g. Appelquist and Balstrøm, 2014; Wilby et al., 2010a). These mainly address impacts on water resources and the coastal zone. One early vulnerability assessment focused on the groundwater resources of Djibouti-Ville, and rural districts of Mouloud and Ali-Sabieh. Hydrological modelling highlighted the effects of reduced infiltration on groundwater levels, and rising salinity of pumped groundwater. Higher sea levels combined with lower rainfall totals further increase the risk of saline intrusion to coastal aquifers (DATE, 2001). Impacts on rural areas include loss of water storage in underground cisterns, loss of pasture and land degradation, and abandonment due to the rising salinity of water used for irrigation.

Djibouti's coastal zone is particularly vulnerable to sea level change. Houssein (2005) tested sensitivity to a rise of 8–39 cm by 2050 combined with a high tide of 60 cm and a storm surge of 120 cm or 180 cm (the 1-in-1000-year event). The combination (i.e. sea level rise plus tide plus surge) yielded water level scenarios of 188 cm and 278 cm (rounded to 2 m and 3 m, respectively). These were then applied to a map of vulnerable sectors in Djibouti-Ville. Under a 2 m high-water level, ~25% of the capital city area would be inundated, affecting ~85,000 inhabitants. Under a 3 m high-water level ~30% of the area is flooded, and ~149,000 inhabitants affected. Furthermore, the Anciens Quartiers of the city would be particularly impacted by back-ponding of sewage during high-water episodes. Dasgupta et al. (2009) estimated that ~60% of the coastal population would be vulnerable to a 10% increase in the severity of the 100-year storm surge. The assessment placed Djibouti amongst the five most exposed developing countries with a coastline (the others being the Bahamas, Kuwait, United Arab Emirates and Belize).

Djibouti-Ville is also subject to the menace of fluvial flooding because the city sits on the alluvial delta of the Oued Ambouli, which lies downstream of a 600 km^2 catchment area. Without adaptation measures, expected changes in the frequency and intensity of storms would lead to a rise in flash flood risk for the city's inhabitants (STDE Consortium, 2007). There is significant exposure to financial losses because of the high concentration of the nation's total domestic, economic and administrative infrastructure in this single floodplain.

According to the ND-GAIN index (Figure 7.3), Djibouti is ranked 152 out of 192 countries in terms of exposure to significant climate change (where a larger number denotes worse conditions). This rank compares with 137 for Eritrea, 143 for Ethiopia and 161 for Somalia. In terms of water security under climate change the rank order of the same neighbouring countries is Djibouti (93), Ethiopia (110), Eritrea (179) and Somalia (181). However, the underpinning data and climate change scenarios are highly uncertain. For example, one study showed that the perennial surface drainage network could decrease by 25% across Africa, but there might be local increases in the east of the continent (de Wit and Stankiewicz, 2006). This is supported by foreseen increases in water resources (e.g. Scholze et al., 2006; Schewe et al., 2014) and lower drought risk (e.g. Prudhomme et al., 2014) with implied gains in surface water availability for some rural communities. Other global assessments suggest that Djibouti could lie within a region of severe water stress by the 2050s (e.g. Alcamo et al., 2007). Such ambiguity amongst outlooks reflects the large uncertainty in regional rainfall scenarios as well as poor representation of semi-arid rainfall–runoff processes in global hydrological models.

Other National Risk Assessments

Djibouti is not alone in undertaking climate risk assessment to shape national adaptation planning. A review of climate change legislation in 99 countries (both rich and poor) found that only one (Libya) had no assessment at all, 50 met the minimum reporting requirements of the UNFCCC, 37 had undertaken a national-level assessment, and 11 had detailed sectoral assessments (Nachmany et al., 2015). The most sophisticated climate risk assessments can be divided into those that are science-led or policy-led initiatives.

Science-led assessments are typically deterministic, beginning with climate change scenarios and ending with anticipated impacts that are used to frame adaptation priorities. The UK's first Climate Change Risk Assessment (CCRA) is a good example of this approach, and highlighted flooding, heatwaves, biodiversity loss and water security as areas of main concern (Figure 9.5). Potential impacts on key sectors were evaluated via a mix of literature review and stakeholder engagement to identify risk metrics for broad-scale quantitative analysis (e.g. Wade et al., 2013). Climatic and socio-economic uncertainties were explored (Chapter 6), with confidence in expected consequences qualified as high, medium or low. The second CCRA took a different approach by focusing more on key vulnerabilities and interconnectedness of climate impacts (Committee on Climate Change, 2016).

Policy-led frameworks begin with adaptation objectives that are deemed to be socially, economically and technically feasible then evaluate their performance under plausible scenarios of

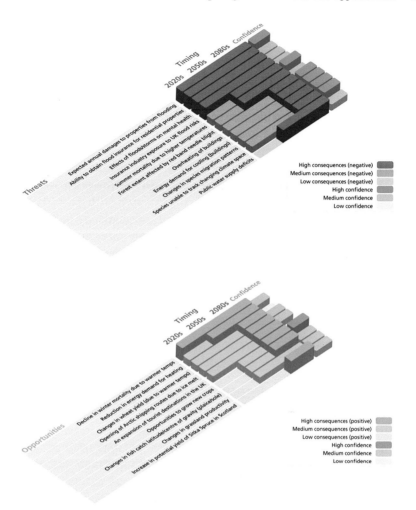

FIGURE 9.5 Threats and opportunities identified by the first UK Climate Change Risk Assessment. *Source*: Defra (2012)

change. This approach was adopted by the Dutch when developing their national strategy for water resources, coastal and fluvial flood risk (Stive et al., 2011). The overall objective of the plan was set out by the Delta Commission (2008:16) as: "*A human life is worth the same everywhere and the probability of a fatality due to a disastrous flood must therefore be assessed on a common basis, to be agreed throughout society.*" Four scenarios of future warming and storminess were developed to test ways of achieving a *probability of fatality* no greater than one in a million. Measures include beach nourishment on a massive scale to defend the coastline against sea level rise and storm surge. The anticipated cost of these works is expected to be €1.2–1.6 billion per annum until 2050.

Case Study Conclusions: Fate of the Francolin

The answer to who or what is most at risk from climate change clearly depends on the unit and scale of assessment. At the global scale, the Horn of Africa is a hotspot of exposure to climate risks, with long-term food and water security the key concerns. Within the region, Djibouti-Ville has high concentrations of people and assets exposed to hazards such as flash flooding and storm surge with potentially disastrous consequences for the national economy. In terms of livelihoods, traditional pastoralists are probably most at risk from more frequent droughts or erratic rainfall. For an individual species, the combination of habitat destruction and drought has already brought the francolin to the edge of extinction. Knowledge of such risks provides the basis for anticipatory action rather than despair.

9.2 GROUP EXERCISE: THREE-STEP CLIMATE RISK ASSESSMENT

NOTES

The purpose of this activity is to reinforce the first three stages in the Willows and Connell (2003) risk assessment framework (Figure 9.3), which were covered by the case study for Djibouti. Recall that the **receptor** of assessment for the exercise may be an *individual* (e.g. university student, lecturer or visitor). The **unit** of assessment may be *place(s)* (e.g. campus, town, county), geographic *domain(s)* (e.g. habitat, upland or coastal zone, built environment), or *sector(s)* (e.g. agriculture, business, energy, environment, health, transport, water).

Set the scene by displaying recent imagery or reports of weather-related impacts on the exemplar unit(s) of assessment. Outline the purpose and scope of the three risk assessment elements: (i) identify problem and objectives; (ii) establish decision-making criteria; (iii) assess risk. Students should work either individually or in small groups to complete the worksheet, pausing after each section to discuss the most significant constraints of the analysis. The group as a whole should then reflect on priorities for further efforts to fill knowledge or data gaps.

Worksheet for a Three-Step Climate Risk Assessment of Specified Unit(s)

Stage 1: Identify problem and objectives Primary objective of the assessment? Present and/or future climate threats? Related non-climatic threats? Policy, programme or project level? Key stakeholders? Time scales for implementing decisions? Short- or long-term benefits?	
Stage 2: Establish decision criteria Measures of decision success? Legislative requirements or constraints? Appetite for risk? Decision-making culture of assessed unit? Ultimate decision-maker? Impact of decision on others? Responsibility held/shared by others? Resource implications? Key receptor(s) and/or exposure unit(s)? Assessment end-point (or payoff)?	
Stage 3: Risk assessment Lifetime of decision or scenario? Most important climate variables? Anticipated impacts of climate variables? Quality of evidence of impacts? Estimated likelihood of impacts? Indirect impacts? Required data and/or tools of analysis?	

Adapted from key questions for Stages 1 to 3 provided by Willows and Connell (2003)

LEARNING OUTCOMES

- To arrive at a deeper understanding of the range of natural hazards faced by urban and rural communities in parts of East Africa.
- To understand the purposes and key features of the Willows and Connell risk and uncertainty framework, and in particular what is involved in implementing the first three stages.
- To recognise that climate risk assessment has to include evaluation of related non-climatic threats linked to socio-economic, demographic and technological development over decision horizons.

9.3 FURTHER QUESTIONS TO RESEARCH AND DISCUSS

What are the essential building blocks of a national climate change risk assessment?

Explain the difference between risk and vulnerability to climate change.

What are the most significant climate threats and opportunities faced by your community?

9.4 FURTHER READING

Petr, M., Boerboom, L.G.J., van der Veen, A. and Ray, D. **2014.** A spatial and temporal drought risk assessment of three major tree species in Britain using probabilistic climate change projections. *Climatic Change,* **124,** 791–803.

Whateley. S., Walker, J.D. and Brown, C. **2015.** A web-based screening model for climate risk to water supply systems in the northeastern United States. *Environmental Modelling and Software,* **73,** 64–75.

9.5 OTHER RESOURCES

Birdlife International on the Djibouti francolin (*Pternistis ochropectus*) www.birdlife.org/datazone/species/factsheet/22678869 [accessed 14/07/16]

Republic of Djibouti National Adaptation Plan of Action 2006 (in French) http://unfccc.int/resource/docs/napa/dji01f.pdf [accessed 14/07/16]

UK Climate Change Risk Assessment 2017 www.theccc.org.uk/tackling-climate-change/preparing-for-climate-change/climate-change-risk-assessment-2017/ [accessed 14/07/16]

United Nations Country Profile for Djibouti http://data.un.org/CountryProfile.aspx?crName=DJIBOUTI [accessed 14/07/16]

US National Climate Assessment 2014 http://nca2014.globalchange.gov/ [accessed 14/07/16]

World Wildlife Fund (2009) notes on the Ethiopian Montane Woodland ecoregion www.worldwildlife.org/ecoregions/at0112 [accessed 14/07/16]

10

How Can Urbanites Avoid Becoming Climate Victims or Villains?

TOPIC SUMMARY

The risk of extreme temperatures in large towns and cities is expected to increase with global warming. More than half of humanity is now urbanised, placing significant numbers of people at risk from future heatwaves that are exacerbated by the heat loading and retaining properties of built environments. Threats to vulnerable groups (such as the young, elderly and those with medical conditions) are further compounded by the poor air quality that often accompanies heatwaves. Although patterns of mortality and morbidity (ill health) typically display a U- or J-shape with temperature, the exact form of the relationship varies between populations. City managers have the challenging task of keeping their citizens safe during heatwaves as well as adapting metropolitan areas to a broader set of climate hazards. Ideally, plans to keep people cool during heatwaves will also yield co-benefits of energy saving, reduced greenhouse gas emissions and improved air quality.

BACKGROUND READING

Sheridan and Allen (2015) assess the consequences of changes in extreme temperatures for human health. Wilby (2007) reviews the potential impacts of climate change on built environments with particular reference to London.

10.1 HEAT IN THE CITY
Silent Yet Growing Threat

A flash flood is a fearful thing to behold. Not the ankle deep nuisance, but the turbulent, mud-laden waves that carry whole trees at the speed of a train. The danger is both immediate and obvious. The same cannot be said for the killer heatwave, which begins as a largely hidden, creeping hazard. Yet, according to the World Meteorological Organisation (2014), extreme temperatures are the deadliest natural disasters affecting North America and Europe (Table 10.1). The European summer heatwave of 2003 claimed 72,210 lives in 15 countries; another summer heatwave in 2010 led to over 55,700 deaths in the Russian Federation alone. The latter

TABLE 10.1 National weather disasters in Europe ranked according to reported deaths. *Source*: World Meteorological Organisation (2014)

Rank	Disaster type	Year	Country	Fatalities
1	Extreme temperature	2010	Russian Federation	55,736
2	Extreme temperature	2003	Italy	20,089
3	Extreme temperature	2003	France	19,490
4	Extreme temperature	2003	Spain	15,090
5	Extreme temperature	2003	Germany	9,355
6	Extreme temperature	2003	Portugal	2,696
7	Extreme temperature	2006	France	1,388
8	Extreme temperature	2003	Belgium	1,175
9	Extreme temperature	2003	Switzerland	1,039
10	Extreme temperature	1987	Greece	1,000

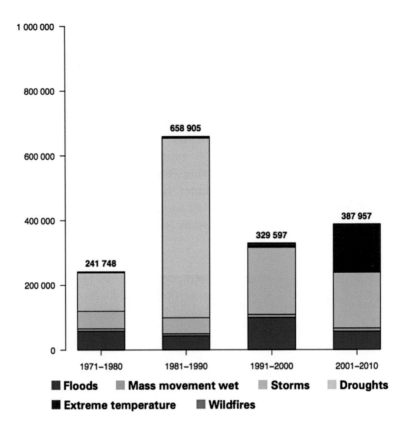

FIGURE 10.1 Number of reported deaths by decade and hazard type for the period 1971–2010. Note that urban and rural data are combined to produce global statistics. *Source*: World Meteorological Organisation (2014)

event was the eighth most lethal of all disasters globally between 1970 and 2012. Overall, the number of reported deaths due to heat stress jumped in the period 2001–2010 (Figure 10.1) and is expected to grow as maximum air temperatures rise over coming decades (Amengual et al., 2014; Lemonsu et al., 2014; Murari et al., 2015).

Higher, more persistent regional temperature extremes are amongst the most certain outcomes of anthropogenic climate change (Kharin et al., 2013). Indeed, several studies assert that past emissions have already increased the risk of heatwaves. For instance, Diffenbaugh and Scherer (2012) claim that a US heatwave similar to that of July 2012 over north-central and northeast regions is now four times more likely compared with pre-industrial conditions. Similarly, Stott et al. (2004) believe that the risk of a European

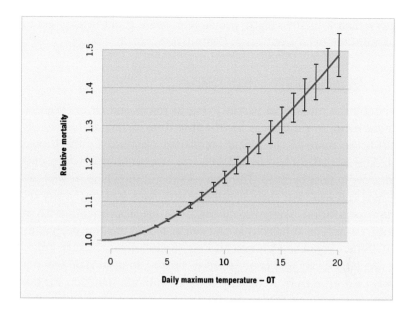

FIGURE 10.2 Relationship between a temperature index (daily maximum temperature minus optimum temperature [OT]) and relative mortality for people aged 65+ years. Reprinted with permission from World Health Organisation (2014) based on data from Honda et al. (2014)

heatwave like that experienced in 2003 may have doubled due to historical GHG emissions. However, rising global temperature extremes have also coincided with a phase of rapid urbanisation since the 1950s. Now there are more people than ever living in built environments that are susceptible to urban warming.

The World Health Organisation (2014) assessed the future global risk of heat-related deaths amongst people aged 65+ using regional temperature projections and a relative mortality–temperature curve (Figure 10.2). This shows the factor by which heat-related deaths increase beyond the optimal temperature (OT) defined as the temperature at which mortality in a population is lowest. Their model suggests a global rise in annual heat-related deaths of 92,207 by 2030, and 255,486 by 2050. The relative increase in excess deaths is expected to be greatest in sub-Saharan Africa, Latin America and south/southeast Asia. The high vulnerability of the last region was highlighted by a heatwave in 2015, when local temperatures reached 48 °C in Allahabad (India) and 49 °C in Larkana (Pakistan), claiming an estimated 4500 lives. By the close of the twenty-first century more than 3 billion elderly people worldwide could be exposed to harmful temperatures (Lancet Commissions, 2015).

An Increasingly Urban Species

In 2009 the number of people living in towns and cities exceeded those living in rural areas for the first time in human history. By 2014, the share of the global population classified by the United Nations as urban had reached 54% – a proportion that is expected to grow to 66% by 2050. The pace of urbanisation is most rapid in Africa and Asia, but these continents still have 90% of the world's rural population. There are currently 28 megacities (each with at least 10 million inhabitants) and home to 453 million people. By 2030 there could be 41 megacities. At 38 million inhabitants, Tokyo is the largest city and is projected to remain so until 2030 when it may be overtaken by Delhi (currently 25 million). Shanghai, Mexico City, Mumbai and São Paulo already have more than 20 million citizens. Such high concentrations of humanity mean that significant numbers of people (and assets) are potentially at risk from weather-related disasters within cities. The urban fabric favours local heat retention whereas the density of economic activity and resource consumption create potent sources of GHG emissions. All three concerns make cities logical units for simultaneously reducing human vulnerability, extreme heat and emissions.

City Rankings by Climate Vulnerability

League tables show ranking of cities by vulnerability (and/or pre-paredness) for climate change. In terms of vulnerability to tropical cyclones and storm surge, globally the most exposed cities are Manila, Karachi, Jakarta, Kolkata and Bangkok (Brecht et al., 2012) (Chapter 7). When considering existing standards of flood protection combined with the impacts of subsidence and sea level rise, the top five most vulnerable cities are Guangzhou (China), Mumbai (India), Kolkata (India), Guayaquil (Ecuador) and Shenzen (China) (Hallegatte et al., 2013). Each could experience average annual losses of US$3 billion by 2050.

Establishing a rank order of city populations most at risk from heatwaves is not straightforward because there are many confounding factors, the most important of which are population age structure, socio-economic status, air pollution and housing

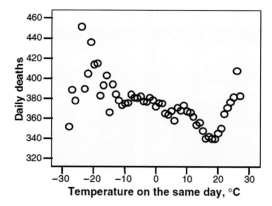

FIGURE 10.3 The daily mortality–temperature relationship for Moscow showing a minimum at 18 °C. *Source*: Revich and Shaposhnikov (2008a)

quality. Other considerations include the prevalence of heat-related cardio-respiratory diseases, public awareness of heat hazards and access to health services. As a consequence, each population has a characteristic threshold above which heat-related mortality rises. This creates a distinctive U- or J-curve of heat mortality versus maximum temperature (Figure 10.3). For example, Moscow and London have an increase of ~3% in non-accidental deaths for every 1 °C beyond the local heat threshold of 18–19 °C (Revich and Shaposhnikov, 2008a; Hajat et al., 2002). By contrast, Monterrey (Mexico) has an increased mortality rate of 19% per 1 °C above the much higher threshold of 31 °C (McMichael et al., 2008). Hence, to rank cities by future risk to extreme temperatures requires knowledge of the expected regional warming in relation to the threshold temperature, any adaptation measures, as well as the size and sensitivity of the population to changes in temperature above that threshold. Within Europe, the citizens of southern cities such as Athens, Marseille, Milan, Rome and Naples, as well as the eastern city of Bucharest, are thought to be at highest risk (Fischer and Schär, 2010).

Urban Hotspots in Space and Time

As indicated before, city-wide statistics may conceal locally high climate vulnerability amongst urban populations (Chapter 7). The most vulnerable groups are the elderly, those with chronic illnesses or pre-existing poor health conditions. The homeless and outdoor

(a)   (b)

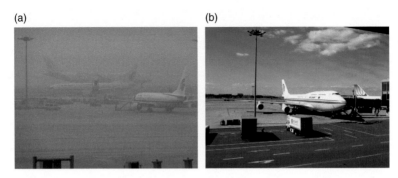

FIGURE 10.4 Air quality at Beijing airport (a) before and (b) after heavy rainfall.

workers also have relatively high exposure to heat compared with the wider population. Poor air quality, especially in China (Figure 10.4) and India, can further contribute to heat-related mortality in ways that exceed simple additive effects. It is estimated that ozone and fine particulates associated with wildfires during the 2010 Russian heatwave contributed an additional 2000 deaths in Moscow amongst those with cardiovascular, respiratory, genitourinary and nervous system diseases (Shaposhnikov et al., 2014). The relative risk of mortality grows with heatwave duration, and in the case of the 2010 Moscow event peaked at about 14 days.

These demographic and socio-economic variations are superimposed on urban landscapes with local hot and cool spots. Warmer areas typically coincide with denser development and thermal loading by artificial heat sources with cooler areas found near green spaces and water features (see Holderness et al., 2013). Urban form and extent also matter: more sprawling US cities tend to have less vegetative cover to protect against solar heating (Stone et al., 2010). The strength of the urban heat island has been observed to fluctuate in some cities, partly because of urban growth and partly due to decadal variations in the frequency of stagnant, high-pressure weather systems that favour high temperatures and poor air quality (e.g. Wilby et al., 2011). Where climate models suggest greater frequency of these atmospheric patterns, there is increased likelihood and intensity of heat island events, which can add several degrees of warming to city centres compared with rural areas (Wilby, 2008). Unsurprisingly, neighbourhoods with the most intense heat islands tend to have higher mortality rates during summer heatwaves (Wong et al., 2013).

Actions to Protect Urban Populations and Cut Emissions

Managers are recognising that the future vitality and economic prosperity of city regions is partly dependent on adapting to climate risks. Weather catastrophes such as Super Typhoon Haiyan and hurricanes Sandy and Katrina have already forced some leaders to reconsider how best to rebuild and plan city life beyond the destruction (Steinberg and Shields, 2008). Various tools are at their disposal, ranging from short-term heatwave forecasts for cities, health care contingency planning, and longer-term strategies to progressively modify building designs, shaded areas and urban layout to better manage extreme temperatures. Other measures include preserving (or expanding) green spaces, greater use of more reflective materials for roads and buildings, or incentivising walking, cycling and travel by mass transit systems. These bring added health benefits from improved air quality and greater physical activity (Stone et al., 2010). Other suggestions include use of river and groundwater within open loop summer cooling (and winter warming) systems for London and Paris. Central Stockholm already makes use of cold water from the Baltic Sea to cool 600 buildings. More radical engineering solutions involve reopening culverted water courses or reconfiguring the urban morphology to improve ventilation and heat advection (Dawson, 2011).

Keeping citizens cool is not the only challenge faced by city managers. There are also climate threats to water and energy supplies, waste disposal and transport systems, river and intense rainfall flooding, air quality, and even biodiversity loss to contend with (Wilby, 2007). Figure 10.5 shows the complex web of dependencies connecting these concerns. Hence, it does not make sense to manage the urban heat island through planning interventions without taking into account transport demands driven by changing population and patterns of employment. Fortunately, there are plenty of win–win options: green spaces provide cool refuges and are generally good for biodiversity and flood management; more energy-efficient buildings and transport systems reduce thermal loads as well as carbon emissions and costs. On the other hand, more compact cities may reduce emissions from transport, but can increase the built mass and propensity for intense heat islands.

Hence, there have been growing calls for a more integrated approach to climate change adaptation and mitigation planning within a broader sustainability framework (Walsh et al., 2011). This

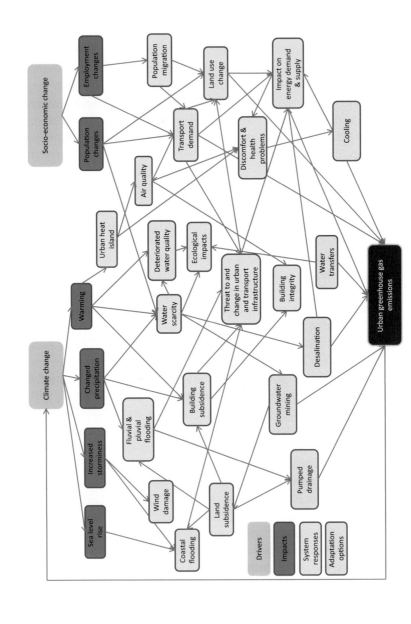

FIGURE 10.5 Some of the complex interactions and interdependencies between climate change, adaptation and mitigation in cities.
Source: Walsh et al. (2011)

involves adopting a city-scale systems approach to developing port-folios of planning policies that reduce climate threats at the same time as curbing emissions across all sectors of the economy. Remote sensing techniques can help by monitoring patterns of heat island development, air quality variations and land cover (e.g. Christen et al., 2011). Using land use as a proxy for urban CO_2 sources and sinks, it is possible to establish carbon budgets at the scale of a metropolitan area (Soegaard, 2003). For example, land surveys and *in situ* CO_2 measurements for Shanghai show highest concentrations (420.5 ± 33.3 ppm) near transportation hubs and those parts of the city with impervious cover at the urban core (Liu et al., 2015a). The International Carbon Action Partnership estimates that the total GHG emissions of Shanghai in 2012 were equivalent to 298 million tonnes of CO_2 [about the same as the entire carbon footprint of Malaysia]. However, investments in more efficient clean energy pro-duction will lower thermal loads and carbon emissions whilst addressing chronic air pollution in Chinese cities (Figure 10.4). According to the WHO, India has now surpassed China as the coun-try with the most polluted cities. During one episode in early 2016, concentrations of ultrafine (PM2.5) particulates in Delhi were 20 times the safe limit (Economist, 2016).

Concluding Remarks

Cities concentrate cultural, technological and economic activity resulting in innovation and wealth; however, if not properly man-aged, cities also amplify climate risks and vulnerabilities (Dawson, 2011:175). This chapter focused on the threats posed by extreme temperatures (and poor air quality) to the future inhabitants of megacities. It is widely assumed that the beneficial effects of rising air temperatures (by reducing winter mortality in cold spells) could outweigh those due to heatwaves in some cities (e.g. London) but this assertion is being challenged by emerging research for other cities. The task facing city managers is how to transition existing energy, water, transport and ecosystem services in ways that improve resilience to extreme hot (and cold) temperatures whilst lowering resource consumption. At the same time, urbanites have a crucial role in carbon management with consequences that extend well beyond city boundaries.

10.2 GROUP EXERCISE: A HEATWAVE PLAN FOR MOSCOW

NOTES

The causes of the 2010 Russian heatwave are not entirely clear. Dole et al. (2011) assert that the event was simply a consequence of natural variability that caused a stubborn area of high pressure over the region. Rahmstorf and Coumou (2011) claim that there is an 80% probability that the observed temperature anomaly (Figure 10.6) could not have occurred without GHG emissions. Otto et al. (2012) find no contradiction between these studies. Their large ensemble of climate model experiments showed that the *severity* of the event could be naturally generated whereas the likelihood of *occurrence* has increased due to past emissions.

The consequences of the heatwave are beyond dispute. Shaposhnikov et al. (2014) report that there were close to 11,000 excess deaths from non-accidental causes during the course of the 44 day heatwave and pollution episode. Earlier heatwaves in 2001 and 2002 claimed an estimated 1200 and 560 lives, respectively, in Moscow alone (Revich et al., 2008b). Hence, it is clear that excess deaths during heatwaves are a serious and recurrent threat to vulnerable groups in Russia.

The aim of this activity is to assemble key components of a heatwave plan for Moscow. Students should prepare independently for the task by reviewing the heatwave plans for the city of Melbourne and the whole of England (following the links provided at the end of this chapter). The most vulnerable groups of Muscovites should first be identified, then specific heatwave actions identified. Finally, students should consider the possibility of integrating the proposed measures within their heatwave plan with actions aimed at reducing other climate risks and GHG emissions. For instance, Moscow is prone to pluvial flooding (as observed during the Great Flood of April 1908, and more recently in July 2012, May 2015 and June 2015). Proposed lists of heatwave measures should be discussed as a group.

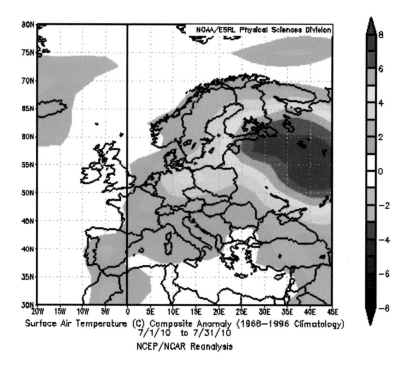

FIGURE 10.6 Temperature anomalies at the height of the record-breaking 2010 Russia heatwave. *Source*: NOAA Earth System Research Laboratory www2.ucar.edu/atmosnews/perspective/2346/how-unusual-moscows-heat

LEARNING OUTCOMES

- To appreciate that heatwaves are the deadliest weather disasters affecting Europe and North America and that global warming combined with a growing/aging urban population increases the likelihood of heat-related mortality.
- To recognise the most important factors shaping the temperature–mortality curve of a city, including variations in socio-economic status, housing quality, public awareness and access to health services.
- To appreciate the co-benefits that can accrue from reconfiguring city landscapes and services in ways that manage heat-related threats alongside other climate hazards whilst reducing carbon emissions.

10.3 FURTHER QUESTIONS TO RESEARCH AND DISCUSS

Explain why the threshold temperature for increased mortality and sensitivity to temperature extremes varies within and between city populations.

What is the so-called 'harvesting effect' and how does this complicate interpretations of mortality during heatwaves?

Give examples of the steps that urban populations could take to both adapt to heatwaves AND reduce the likelihood of extreme temperatures in the future.

10.4 FURTHER READING

Kinney, P.L., Schwartz, J., Pascal, M., et al. 2015. Winter season mortality: will climate bring benefits? *Environmental Research Letters*, **10**, 064016.

Trenberth, K.E. and Fasullo, J.T. 2012. Climate extremes and climate change: the Russian heat wave and other extremes of 2010. *Journal of Geophysical Research – Atmospheres*, **117**, D17103.

10.5 OTHER RESOURCES

C40 Cities www.c40.org/ [accessed 14/07/16]

COP21: City mayors discuss green solutions www.bbc.co.uk/news/science-environment-34999962 [accessed 14/07/16]

City of Melbourne Personal Heatwave Response Plan www.melbourne.vic.gov.au/community/safety-emergency/emergency-management/Pages/Heatwaves.aspx [accessed 14/07/16]

Climate change risks to London www.youtube.com/watch?v=y12rlCBRc4w [accessed 14/07/16]

Heatwave plan for England www.gov.uk/government/uploads/system/uploads/attachment_data/file/429384/Heatwave_Main_Plan_2015.pdf [accessed 14/07/16]

Oral history of the 1995 Chicago heatwave www.chicagomag.com/Chicago-Magazine/July-2015/1995-Chicago-heat-wave/ [accessed 14/07/16]

Plan National Canicule (heatwave advice in French)
www.sante.gouv.fr/canicule-et-chaleurs-extremes.html [accessed 14/07/16]

Where is the world's hottest city? www.theguardian.com/cities/2015/ jul/22/where-world-hottest-city-kuwait-karachi-ahvaz [accessed 14/07/16]

11

What Is Dangerous Climate Change?

TOPIC SUMMARY

The global mean surface temperature (GMST) has risen by 1 °C since pre-industrial times. Further warming carries growing risks of adverse consequences for human and natural systems. However, defining a threshold temperature for *dangerous* climate change is as much a value judgement as technical challenge. If worth is attached to safeguarding coral reefs then the limit of global warming might be less than 1 °C; if avoiding a collapse of the West Antarctic Ice Sheet then the threshold could be between + 3 °C and 5 °C. The temperature target, in turn, determines the allowable CO_2 concentration and with it the differentiated responsibilities of nations to cut emissions. The widely accepted global warming target of +2 °C compared with pre-industrial temperatures translates into cumulative CO_2 emissions of about one trillion tonnes (1000 GtC). Some climate models suggest that this temperature change could be reached by the 2050s without drastic reductions in emissions.

BACKGROUND READING

Schellnhuber et al. (2006) document the science presented at a seminal conference on *Avoiding Dangerous Climate Change*. This body of research was a backdrop to the G8 Gleneagles Summit at which world leaders agreed to make substantial cuts in GHG emissions.

11.1 PROGNOSIS, SYMPTOMS AND SAFE DOSE
Safety Limits for People and Planet

Health and safety limits are meant to protect whilst being 'reasonably practical'. For instance, legal standards offer defence against noise in the work-place, radioactivity in the environment, or trace metals in our food. Other safe limits are given by public health authorities for alcohol and processed meat consumption. These are intended to promote healthier lifestyles or to deter harmful behaviours. Such advice varies between countries and is complicated by a host of factors, not least age and gender. For instance, it has been estimated that a 20-year-old man who drinks 4 units of alcohol per day might expect to lose about 20 seconds of life for every pint consumed (Spiegelhalter and Riesche, 2008). More alarming, the US Surgeon General advises that there is no safe level of cigarette smoking. In this case, a 30-year-old smoker reduces his life expectancy by 10 years if the habitat is not broken (Doll et al., 2004). This equates to a staggering 6 hours of future life lost for every day spent smoking (Spiegelhalter and Riesche, 2008).

Similar thinking lies behind international efforts to avoid dangerous climate change. As with the heavy drinker, setting a 'safe limit' for global CO_2 emissions involves working out (i) the costs and benefits of no action, (ii) symptoms or danger signs, and (iii) the dose–response relationship. For the drinker, a safe level might reflect (i) improved life expectancy and financial savings weighed against pleasures foregone, (ii) incidence of cancer, weight gain or loss of cognitive function, and (iii) recommended limits for alcohol intake beyond which these harms escalate. For the global climate, the allowable emissions might reflect (i) avoided damage to ecosystems, food production and economic development, (ii) tipping points in Earth systems that could lead to irreversible or dramatic changes in climate, and (iii) the amount of GHGs that could be emitted without exceeding a specified global mean temperature or sea level rise.

After sketching the history of evolving definitions of dangerous climate change, subsequent sections expand on each of the three elements outlined above (i.e. prognosis, symptoms and safe limits). The last section briefly touches on the complex issue of assigning liabilities and, by implication, responsibility for the costs of avoiding dangerous climate change.

Evolving Concepts of Dangerous Climate Change

Civilisations have always had a healthy respect for the potential threats posed by climate, even if historical fears about the impending return of ice ages are now regarded as misplaced (Chapter 2). Over 50 years ago, the US President's Science Advisory Committee (1965) warned Lyndon Johnson that the greenhouse effect would "*almost certainly cause significant changes*" and "*could be deleterious from the point of view of human beings*". However, first use of the term 'dangerous climate change' is accredited to the United Nations Framework Convention on Climate Change (UNFCCC). Article 2 of the UNFCCC (1992) stated that the ultimate objective of the Convention is "*stabilization of greenhouse gas concentrations in the atmosphere at a level that would prevent **dangerous** anthropogenic interference with the climate system. Such a level should be achieved within a time frame sufficient to allow ecosystems to adapt naturally to climate change, to ensure that food production is not threatened and to enable economic development to proceed in a sustainable manner.*"

Not long afterwards, Moss (1995:4) perceived that what constitutes dangerous interference with the climate system is a value judgement. Scientific research can explore the dose–response relationship between GHG emissions, climate variables, human and physical system functions; but the goals set for protecting those vulnerable systems are determined by socio-political priorities. Science is not well placed to answer the question: What *should* the atmospheric concentration of CO_2 be to achieve an optimal climate state for humans?

The IPCC (2001) Third Assessment Report tackled this conundrum by relating global mean temperature increases to five 'reasons for concern'. These are: (i) damage/irreparable loss of unique and threatened natural systems (e.g. coral reefs); (ii) greater probability of extreme events (e.g. droughts and floods); (iii) uneven distribution of impacts (e.g. increased crop yields in some areas but decreases in others); (iv) growing aggregate worldwide damages (e.g. from sea level rise and coastal flooding); and (v) increased likelihood of large-scale singular events (e.g. disintegration of the West Antarctic Ice Sheet, or shutdown of the North Atlantic Thermohaline Circulation [THC]). These concerns were represented by the so-called 'glowing embers' diagram (Figure 11.1).

Article 3 of the UNFCCC speaks of "*serious and irreversible damages*" and the need for "*precautionary measures to anticipate, prevent or*

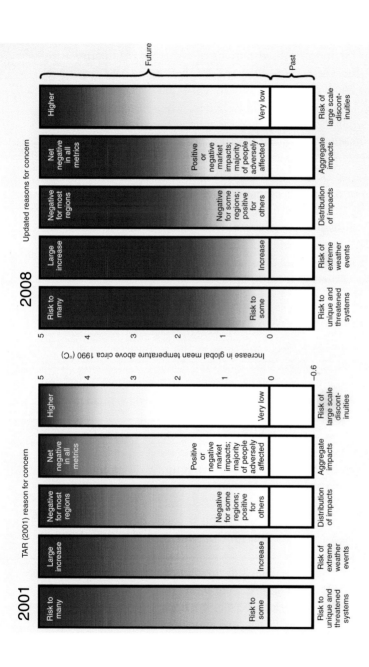

FIGURE 11.1 Impacts of, or risks from, climate change, by reason for concern. Each column corresponds to a reason for concern, and shades indicate severity of impact or risk. White means no or virtually neutral impact or risk, yellow means somewhat negative impacts or low risks, and red means more negative impacts or higher risks. The left panel shows the original assessment by IPCC (2001), which was revisited in 2008 (right panel) ahead of the Copenhagen summit. Differences between the two panels reflect advances in the scientific understanding of global warming impacts. Sources: IPCC (2001), Smith et al. (2009) and Anderson (2012).

minimise the causes of climate change and mitigate its adverse effects". The principle of "*differentiated responsibilities and respective capabilities*" amongst nations was also established as the basis for combating climate change. The 1997 Kyoto Protocol set the modest (but achievable goal) for industrialised nations of reducing cumulative emissions for the years 2008–2012 to ~5% below the 1990 emissions levels. In 2007 the European Council stated: "*Climate change is happening. Urgent action is required to limit it to a manageable level. The EU must adopt the necessary domestic measures and take the lead internationally to ensure that global average temperature increases do not exceed pre-industrial levels by more than 2 °C.*" The 2 °C limit was also accepted by signatories of the 2009 Copenhagen Accord. In this way the definition of dangerous climate change as +2 °C became embedded within a legal framework that ultimately led to the creation of the European Emissions Trading Scheme (Chapter 13). In 2015, the United Nations Conference of the Parties (COP) acknowledged that dangerous impacts could occur at lower levels of global warming and shifted discussions to a 1.5 °C limit.

Prognosis for a World with Business as Usual

Scientific evidence is accumulating on the serious regional consequences of unmitigated GHG emissions (Figure 11.1). Some work begins with the premise that trajectories will overshoot +2 °C and are actually on track to a +4 °C world marked by widespread impacts from sea level rise, extreme heatwaves, food and water insecurity (e.g. World Bank, 2013). Depending on the response of the Greenland and West Antarctic ice sheets, global mean sea level could rise by as much as 2 m by 2100, thereby displacing up to 187 million coastal inhabitants over the twenty-first century (Nicholls et al., 2011). In a +4 °C world, 74% of major river basins could become more water stressed than at present (although this is partly due to rising water demand linked to population growth over the same period) (Fung et al., 2011).

   Concerning risks to unique and threatened ecosystems, one early study suggested that widespread coral reef demise may be unavoidable because of historical GHG emissions and the associated warming committed by the thermal absorption and inertia of oceans (Keller et al., 2005). Corals and reefs are already responding to rising water temperatures by species replacement and bleaching. Data for hundreds of colonies across the Great Barrier Reef (GBR), off the

Australian coast, show that calcification and growth of coral has declined by more than 13% since 1990 – a response that is unprecedented in the last 400 years. Such losses are attributed to the dual stresses of higher water temperatures and declining saturation of aragonite in seawater (De'ath et al., 2009). Others are more optimistic that the waning of the GBR may be reversed by better governance and management of coral reefs (Hughes et al., 2015).

Risks from extreme events include more severe and widespread droughts, flooding and storm damage, as well as greater variability between these extremes (see Figure 10.1). For example, climate models suggest greater aridity as a consequence of both decreased precipitation and increased evaporation over densely populated regions such as Southern Europe, the eastern United States, Southeast Asia and Brazil (Dai, 2013). Under Representative Concentration Pathway (RCP) 8.5 some climate models project drought covering more than 40% of the land area by the end of the twenty-first century (Prudhomme et al., 2014). In addition, one assessment of global flood risk by 2050 under the high emissions SRES A1b scenario showed that the frequency of the current 100-year flood could double across 40% of the globe, with attendant damages rising by ~187% compared with a baseline of no climate change (Arnell and Gosling, 2016).

The distribution of unmitigated climate change impacts is expected to be uneven, with significant negative consequences for some regions and benefits for others. This outcome is reflected by the experiments mentioned previously for drought risk (Figure 11.2). According to these model projections the agro-economy of the Mediterranean basin could be a major loser, whereas reduced aridity over the Lake Victoria Basin, East Africa, could favour increased productivity. Likewise, the global assessment of future flooding found that risk increases for some populous areas of south and east Asia, but decreases over large swathes of central Europe (Figure 11.3). However, both studies stress that the uncertainty in projected risks between climate models is greater than the range amongst different emissions and socio-economic scenarios. This highlights the major technical challenge of scaling regional impacts with rises in global mean temperature (as in Figure 11.1).

Aggregate (i.e. net) impacts recognise that some communities benefit from, for example, less flooding, whereas others are more at risk. When benefits are outweighed by negative consequences, the aggregate impact becomes more severe with global warming

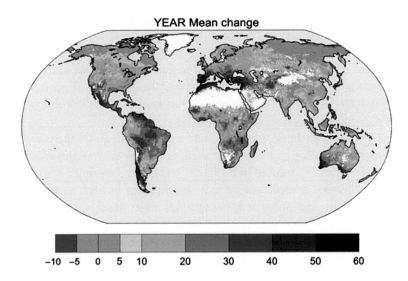

FIGURE 11.2 Percentage change in the occurrence of days under drought conditions for the period 2070–2099 relative to 1976–2005, based on an ensemble of five GCMs driving seven global hydrological models under RCP8.5. *Source*: Prudhomme, C., Giuntoli, I., Robinson, E.L., et al. 2014. Hydrological droughts in the 21st century, hotspots and uncertainties from a global multi model ensemble experiment. Proceedings of the National Academy of Sciences of the United States of America, 111, 3262–3267.

(the middle 'ember' in Figure 11.1). For example, Arnell and Gosling (2016) found that approximately 450 million people and 430,000 km^2 of cropland could be exposed to higher flood risk by 2050. At the same time about 75 million people and 180,000 km^2 of cropland could benefit from reduced flood risk. Hence, the aggregate climate change impact in this case (375 million people and 250,000 km^2) is negative. One example of a positive aggregate impact of global warming might be fewer cold-related deaths in some regions (Field et al., 2014).

Risks from future large-scale discontinuities may include disintegration of the Greenland and West Antarctic ice sheets, disappearance of Arctic summer sea ice, or slowdown of the THC due to a massive influx of Greenland meltwater. These non-linear responses become more likely under higher rates of global warming and once triggered could lead to irreversible changes in Earth systems. For example, Solomon et al. (2009) found that if CO_2 concentrations peak at 1200 ppm (roughly three times the 2015 level) and emissions are

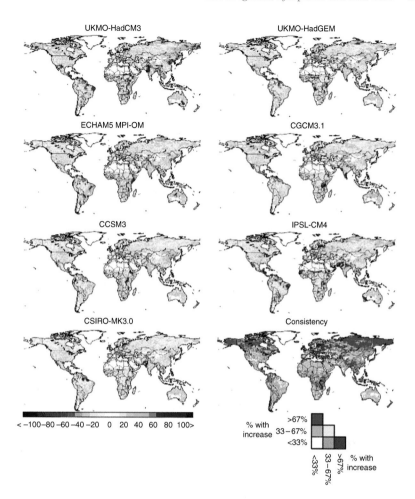

FIGURE 11.3 Percentage change in the magnitude of the estimated 100-year flood under SRES A1b emissions by 2050, for seven climate models, plus the consistency (expressed as a percentage of the total number of models) in projected change across 21 climate models. Areas where baseline average annual runoff is less than 10 mm/yr are shaded grey. *Source:* Arnell and Gosling (2016)

cut to zero afterwards, the world could still be committed to at least 4 °C warming and 2 m of thermal expansion with sea levels rising beyond year 3000. At these higher temperatures, large releases of CO_2 from soils and forest die-back plus methane from wetlands, ocean sediments and thawing permafrost are more likely to increase.

These add to the burden of atmospheric GHGs by creating a positive feedback of more warming and more terrestrial emissions.

Symptoms of Danger

There are likely to be warning signs of such large-scale changes in Earth systems. Considerable effort has been invested in defining both the immediacy and thresholds for dangerous climate change. Table 11.1 gives examples of indicators that have been used to define danger using physical and social vulnerability criteria. Several have already been mentioned before, but the technical challenge is how to *measure* these vulnerable elements. For example, what data would be used to monitor the strength of THC ahead of a complete breakdown? In this case, the volume of water flowing through a transatlantic section at 26° N and at a depth of about 1000 m is widely used; but records only began in the 1950s (Bryden et al., 2005) and were piecemeal until the RAPID array was established (Wells, 2016).

Assessing dangerous changes in terrestrial systems can be just as problematic. Key considerations here are measuring the amount of inertia in the system (i.e. how quickly does it respond to external

---

TABLE 11.1 Examples of physical and social indicators of dangerous climate change. Adapted from Dessai et al. (2004)

Danger measured through thresholds in physical vulnerability

- Large-scale eradication or degradation of coral reef systems
- Disintegration of the West Antarctic Ice Sheet
- Breakdown of the North Atlantic THC
- Changes in the frequency and strength of modes of variability such as El Niño
- Rate of climate change exceeds pace at which species/biomes can migrate

Danger measured through thresholds in social vulnerability

- Regional irrigation demands exceed sustainable water resource supplies
- Depopulation of sovereign atoll countries
- Number of people at risk from water scarcity, hunger, malaria and flooding
- Destabilisation by environmental conflicts and refugees
- Economic impacts of weather disasters exceed a threshold percentage of GDP

pressures) and the amount of variability that is intrinsic to the system (i.e. with background 'noise' obscuring detectable changes) (Contamin and Ellison, 2009). Potentially useful but still complex indicators might include changes in phenology (the seasonal behaviour of species), geographic range (such as upward migration of treelines), community composition (via extinctions or entry of new species), and altered dynamics between species (such as wolf predation and moose herbivory) (Walther et al., 2002).

Social vulnerability indicators of danger (Table 11.1) are shaped by individuals' experienced or perceived levels of insecurity (Dessai et al., 2004). This means that what constitutes danger depends on psychological, social, moral, economic, institutional and cultural factors. [See the web-link to 'Insane rope-swing' as an extreme case of danger-denial.] The legitimacy and amount of information available on physical vulnerabilities also matter. Sometimes a prediction of danger can be self-fulfilling. For instance, depopulation of an atoll might be hastened because finite groundwater is exhausted when residents foresee that sea level rise will contaminate the source and destroy their homes eventually (Dessai et al., 2004).

Allowable Global Emissions

With an appreciation of the dangers ahead, and the means to track symptoms of climate change using physical and social vulnerability indicators, one key question remains: What is a 'safe' amount of carbon emissions to the atmosphere? The answer depends on (i) the agreed outcome or definition of safe and (ii) the sensitivity of the outcome to carbon emissions (i.e. dose–response). Knutti et al. (2016) contend that no scientific assessment has yet defended the widely assumed goal of +2 °C and that it is unclear what level can be considered safe.

If safe means avoiding current widespread bleaching of corals then it may already be too late. If halting irreversible melt of the land-based Greenland Ice Sheet (and long-term commitment to up to 7 m of sea level rise) then a global mean temperature change of +2 °C may be the limit. Hopefully, die-back of the Amazonian rainforest may be avoided below +3–4 °C, and shutdown of the Atlantic THC at +3–5 °C (Lenton et al., 2008). Other so-called 'tipping points' exist for loss of Arctic summer sea ice (+0.5–2 °C), disruption of El Niño (+3–6 °C) and collapse of the West Antarctic Ice Sheet with up to 5 m of sea level rise (+3–5 °C). Note, however, that the latter may be preceded by disintegration of ice shelves and acceleration of ice flows to the ocean.

Taking 2 °C as the safety line, it has been estimated that there is a 50% chance of avoiding dangerous climate change if future cumulative carbon emissions are limited to 1000 GtC (Allen et al., 2009). This temperature would certainly be exceeded if all recoverable fossil fuels are burnt (Meinshaussen et al., 2009). To put these figures into context, the Global Carbon Project estimates that total cumulative carbon emissions from 1870 to 2014 were 545 ± 55 GtC (of which 400 ± 20 GtC were from fossil fuels and cement, and 145 ± 50 GtC from land use change, deforestation and cattle ranching). This means that we could be over half-way to the 1000 GtC target and now adding carbon to the atmosphere at a rate of ~10 GtC per year. Or, to put it another way, at the present level of emissions there is an even chance of global warming reaching +2 °C by the 2050s. However, some very pessimistic model experiments suggest that dangerous climate change could occur at much lower cumulative emissions. For example, Zickfeld and Bruckner (2008) found that the THC could shut down at emissions totalling just 870 GtC. Optimists believe that wind patterns rather than salinity drive the world's ocean currents and that the THC is unlikely to shut down in the foreseeable future (see O'Hare, 2011).

Figure 11.4 shows the relationship between cumulative CO_2 emissions and global mean temperature change. As well as the near-linear scaling of temperature with emissions, the graph also shows the effect of different emission pathways. Under RCP2.6, the climate model ensemble mean is less than +2 °C even by 2100, but under RCP8.5 this warming is reached by the 2050s (see also Figure 6.4). The spread of the shading represents the uncertainty due to climate model sensitivity and natural variability (Chapter 4). This means that there is a small chance that +2 °C could occur sooner than the 2050s, regardless of the emissions scenario. Even worse, attention to only the global mean temperature change underplays the much larger regional climate signals and impacts. For example, the change in coldest night-time temperature has already surpassed +2 °C in the Arctic, and a change in hottest day-time temperature could exceed +2 °C around the Mediterranean by 2030 and in the contiguous United States by the mid 2040s (Seneviratne et al., 2016).

When allowable CO_2 emissions are estimated with regional impacts in mind, the cumulative totals are lower than for global mean temperature targets. This is because changes in regional extreme temperatures over land are much larger than changes in the overall global (ocean and land) surface mean. Hence, to limit warming of hot

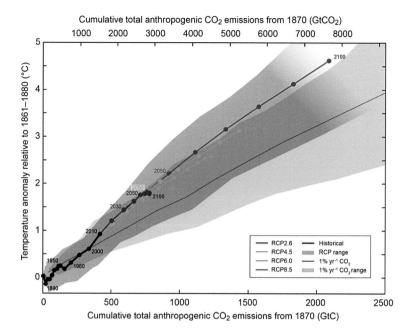

FIGURE 11.4 Global mean surface temperature increase as a function of cumulative total global CO_2 emissions from various lines of evidence. Multi-model results from a hierarchy of climate–carbon cycle models for each RCP until 2100 are shown with coloured lines and decadal means (dots). Some decadal means are indicated for clarity (e.g. 2050 indicating the decade 2041–2050). Model results over the historical period (1860–2010) are indicated in black. The coloured plume illustrates the multi-model spread over the four RCP scenarios and fades with the decreasing number of available models in RCP8.5. The multi-model mean and range simulated by CMIP5 models, forced by a CO_2 increase of 1% per year (1% per year CO_2 simulations), is given by the thin black line and grey area. For a specific amount of cumulative CO_2 emissions, the 1% per year CO_2 simulations exhibit less warming than those driven by RCPs, which include additional non-CO_2 drivers. All values are given relative to the 1861–1880 base period. Decadal averages are connected by straight lines. Source: IPCC (2013a), Figure SPM.10

extremes in the Mediterranean to +2 °C implies a global mean of +1.4 °C which, in turn, translates into cumulative emissions less than 600 GtC (Figure 11.5). The same reasoning can be applied to other extremes that scale approximately with global mean temperature. For instance, to limit changes in 5-day heavy precipitation events over South Asia to less than 10%, the model mean suggests cumulative emissions less than ~700 GtC (Seneviratne et al., 2016).

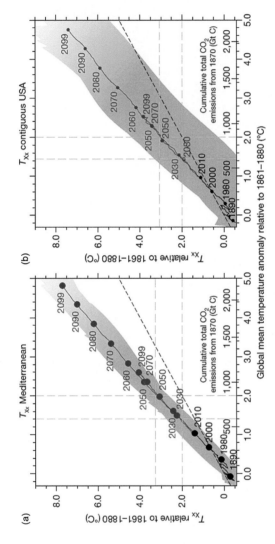

FIGURE 11.5 Scaling between regional changes in annual hot-day extremes (T_{Xx}) and changes in global mean temperature, with associated global cumulative CO_2 emissions targets for (a) the Mediterranean and (b) the contiguous United States. Solid blue (red) lines denote the climate ensemble mean under RCP4.5 (RCP8.5); shaded areas are the climate model range; black dashed lines are the 1:1 scale for global to regional temperatures; grey dashed lines show the temperatures or CO_2 emissions associated with 2 °C increases in global mean and regional extreme temperatures, respectively. Source: Seneviratne et al. (2016)

Uneven Consequences and Responsibilities

Finally, the mismatch between national responsibilities for historical CO_2 emissions and the expected distribution of regional climate impacts (e.g. Figures 11.2 and 11.3) brings to mind the language of *"differentiated responsibilities"* in Article 3 of the UNFCCC. The possibility of legal action has been contemplated by those developing real-time climate attribution systems (Allen, 2003). Within the realm of modelled worlds, the change in risk can be attributed to historical emissions by comparing with no emissions experiments (just as an increased risk of cancer can be compared between groups of smokers and non-smokers). Thus far, lawsuits are filed typically either with the intention of blocking polluting projects or to recover claims for personal injury, property or economic damages allegedly attributed to climate change. One of the most notable cases was filed in 2007 by the state and local governments of Massachusetts, who pressed the US Environmental Protection Agency to regulate CO_2 emissions under the federal Clean Air Act. Despite these developments, Funke (2010) argues that *"climate change is a case for regulation rather than litigation"* because most GHG emissions are linked to basic consumer needs (such as food, transport and energy). In this respect, everyone is both a potential litigant and claimant.

11.2 GROUP EXERCISE: NEGOTIATING POSITIONS AT PARIS CONFERENCE OF THE PARTIES

NOTES

Intended Nationally Determined Contributions (INDCs) mark a departure from the top-down approach to setting emission targets under the Kyoto Protocol. The new bottom-up framework allows nation states to set out their own plans to reduce CO_2 emissions and adapt to climate change by 2030. By the eve of the 21st Conference of the Parties (COP 21) held in Paris in December 2015, 184 countries covering ~95% of global GHG emissions had submitted their INDCs. These documents vary in their detail and content but they do shed light on national priorities and negotiating positions.

This activity involves reviewing a contrasting set of INDCs from the UNFCCC website (see link under Other Resources). Six

cont.

countries/negotiating blocks are suggested (Brazil, China, Ethiopia, EU, Maldives and United States) to represent a range of perspectives, but others might be chosen (e.g. Saudi Arabia as a major fossil fuel exporter, or Eritrea as a highly vulnerable and poorly prepared country for climate change, see Figures 7.4 and 7.5). When completing the worksheet, students should capture key features of the selected INDCs. In particular, they should note the various types of target and base year, any conditions attached to the target, range of measures planned, and evidence provided on fairness and ambition.

Negotiating Positions at Paris Conference of the Parties (COP 21)

Worksheet			
Unit	CO_2 target (2030)	Measures	Fairness and ambition
Brazil			
China			
EU			
Ethiopia			
Maldives			
United States			

LEARNING OUTCOMES

- To be aware of the various scientific and social factors shaping definitions of dangerous climate change.
- To recognise that the widely accepted global mean temperature target of +2 °C implies much larger changes in regional temperature (and attendant impacts) for most land areas.
- To appreciate that allowable targets for CO_2 depend on the target (regional or global mean) temperature and climate sensitivity to CO_2.

11.3 FURTHER QUESTIONS TO RESEARCH AND DISCUSS

What is the likelihood of limiting global mean warming to less than +2 °C? What are the potential implications of over-shooting this target?

Explain why a global mean temperature target of +2 °C is not sufficiently stringent when viewed from a regional impact perspective.

Account for differences between individuals' perceptions of dangerous climate change.

11.4 FURTHER READING

Davidson, E.A. and Janssens, I.A. 2006. Temperature sensitivity of soil carbon decomposition and feedbacks to climate change. *Nature*, **440**, 165–173.

Oppenheimer, M. and Petsonk, A. 2005. Article 2 of the UNFCCC: historical origins, recent interpretations. *Climatic Change*, **73**, 195–226.

Ricke, K.L., Moreno-Cruz, J.B., Schewe, J., Levermann, A. and Caldeira, K. 2016. Policy thresholds in mitigation. *Nature Geoscience*, **9**, 5–6.

11.5 OTHER RESOURCES

Intended Nationally Determined Contributions (INDCs) www4.unfccc.int/submissions/indc/Submission%20Pages/submissions.aspx [accessed 15/07/16]

RAPID: Monitoring the Atlantic Meridional Overturning Circulation at 26.5° N www.rapid.ac.uk/rapidmoc/ [accessed 15/07/16]

Reducing Emissions from Deforestation and Forest Degradation (REDD+) www.un-redd.org/aboutredd [accessed 15/07/16]

The Global Carbon Project www.globalcarbonproject.org/index.htm [accessed 15/07/16]

'The most insane rope-swing bungee jump. Ever' – video www.theguardian.com/sport/video/2014/mar/14/rope-swing-bungee-jump-magwa-falls-south-africa-video [accessed 15/07/16]

12

Why and How Are Carbon Footprints Measured?

TOPIC SUMMARY

A carbon footprint (CF) is defined as the total CO_2 emissions that are both directly and indirectly accrued by an activity or product life-cycle. The carbon may be allocated to different units of assessment, such as nation states, sectors or individuals, using data from GHG inventories. Such statistics inform carbon management and mitigation plans by highlighting 'hotspots' for energy efficiency gains or changes in behaviour. The data may also be used as a comparator or measure of compliance with targets for CO_2. Energy (carbon) savings are increasingly set alongside the conservation of other natural resources such as food and water to achieve multiple benefits. Carbon footprints can be calculated for emissions embodied in international commodities and services as well as for whole cities.

BACKGROUND READING

Pandey et al. (2011) provide an overview of CF guidelines and methodologies used by governments, corporations and the general public. Zhang et al. (2015) demonstrate the power of product life-cycle assessment in identifying carbon-intensive stages of the humble T-shirt, from cotton cultivation to clothing disposal.

12.1 TO MEASURE IS TO MANAGE
How's That?

Cricket is a puzzling game. Apparently, there are five common and five obscure ways that a batsman can be declared 'out'. The most often ones are when: (i) the wicket is struck by the ball (bowled); (ii) the ball is struck then caught by a fielder; (iii) the batsman's leg is before the wicket (so blocking an on-target ball); (iv) either batsmen fail to reach the opposite end of the pitch before the ball strikes a wicket (run out); or (v) the batsman steps outside the crease (perhaps when aiming for a ball) and the wicket is struck by a ball thrown by any fielder. The rarer ways to be out are when: (vi) the batsman accidently strikes his own wicket; (vii) or handles the ball without permission from the fielding side; (viii) the ball is hit twice; (ix) the batsman obstructs a fielder from making a catch or run out; and (x) a new batsman takes too long to enter the field of play (timed out).

Carbon accounting is the process by which the amount of CO_2 emitted by an entity is calculated. As will be shown below, it is also a mechanism for creating carbon credits as a commodity to be traded on markets. Like cricket, carbon accounting has a singular purpose but many different ways of achieving the desired outcome. In this case, the 'game plan' is to drive down global emissions of CO_2 by assigning the weight of responsibility to emitters. Using another metaphor – there is more than one way to divide a cake. By 2015, the size of that (global carbon) cake was growing by ~10 GtC/yr (equivalent to 35 billion tonnes of CO_2) on top of the 545 GtC that had already been emitted since the start of the Industrial Revolution (Chapter 11).

This chapter describes several common and some less obvious units of assessment for characterising carbon emissions (i.e. the carbon 'footprint'). The shared thinking is that for carbon emissions to be managed, they have to be measured in consistent and transparent ways. Such data are essential to the development of plans to avoid dangerous climate change (Chapter 11) by decarbonising economies (Chapter 13).

National Footprints

League tables for national CO_2 emissions are perhaps the most familiar type of CF. These are constructed following internationally agreed standards for determining boundaries for emissions (and carbon removal), and by compiling detailed inventories of all anthropogenic sources of GHG emissions from four main sectors: (i) agriculture/ forestry/land use, (ii) energy, (iii) industrial processes and products, and (iv) waste (IPCC, 2006). Tiers (or scopes) are used to allocate emissions to different boundaries. Tier 1 includes all direct emissions that occur from sources owned or controlled by the assessed entity (e.g. emissions from combustion in boilers, furnaces or vehicles). Tier 2 covers emissions from the generation of energy purchased by the entity (e.g. off-site electricity supply for an industrial process). Tier 3 includes all other indirect emissions that are a consequence of the activities of the entity, but occur from sources not owned or controlled by it (e.g. business travel and procurement).

The radiative forcing by different GHGs is expressed in the common currency of the global warming potential (GWP), benchmarked to the CO_2 equivalent (CO_2-eq). The GWP is defined as the amount of heat trapped by a unit mass of gas. This depends on the infrared radiation absorption properties and atmospheric lifetime of the gas compared with CO_2. For example, the GWP of methane (including indirect effects of tropospheric ozone and stratospheric water vapour production) is 56 over 20 years and 21 over 100 years (compared with 1 unit for CO_2). The GWP for sulphur hexafluoride is a staggering 16,300 units over 20 years.

Variations in audit periods and data quality can make comparisons over time and between countries problematic. National emissions statistics may also differ from those based on international data. This is because assumptions have to be made when converting between units such as weight of coal (tonnes) turned into energy (10^{12} joules or one terajoule [TJ]) and emission factors (e.g. kg CO_2 per TJ coal combusted). Differences may also arise due to discrepancies in estimated fossil-fuel statistics, or in the assumed carbon efficiency at the scale of individual energy plants. Even the type of coal matters, with increasing hardness grades spanning peat (a coal precursor), lignite (or brown coal), bituminous coal (widely used for electric power generation) and anthracite (for residential and commercial space heating). For instance, the energy content (and carbon percentage) by mass is ~28,000 kJ/kg (60–75%) for lignite compared with ~35,000 kJ/kg (>90%) for anthracite. With all this complexity for a single fuel (coal) and sector (energy), imagine the challenge when estimating economy-wide emissions.

Historical National Emissions

Greenhouse gas emissions and removals data compiled by the EU Joint Research Committee show the top three national emitters of CO_2 were China (10.54 billion tonnes), the United States (5.33 billion tonnes) and India (2.34 billion tonnes) (Olivier et al., 2015). These three nations alone accounted for ~53% of total global CO_2 emissions. The 28 member states of the European Union (EU28) added a further 3.42 billion tonnes (or ~10% to the global total). At the other end of the spectrum were the US Virgin Islands, Nauru and Tuvalu with just about 1 kilotonne each. Whereas the CO_2 emissions of the United States and EU28 have declined since the 1990s, China accounts for most of the global increase over the same period (Figure 12.1). Up until 2009, China and India had contributed less than 11% of cumulative emissions since 1750 – nearly the same as the total historical carbon of Germany plus the UK (13%) over the same period. The United States had by then already emitted 28% of the total CO_2 burden from humanity (Monastersky, 2009).

Clearly, population size is a major factor driving national CFs. In 2016, the combined population of China, the United States and India was over 3 billion people (~41% world share), compared with 125,000 people (~0.002% world share) for the US Virgin Islands, Nauru plus Tuvalu. Even so, the proportion of global CO_2 emissions is still greater for the top three than would be expected given their aggregate population. This is explained by a host of other factors, amongst which the economic structure and fuel mix are most important. In

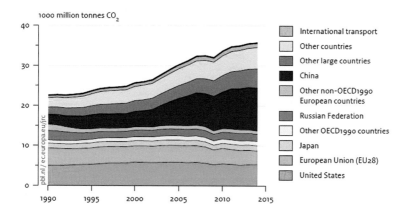

FIGURE 12.1 Global CO_2 emissions per region from fossil-fuel use, international transport and cement production. *Source*: Olivier et al. (2015)

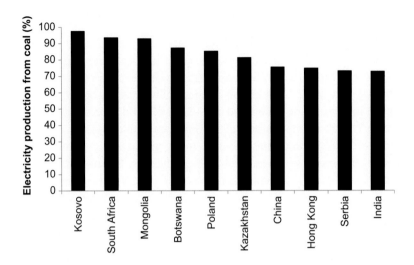

FIGURE 12.2 The ten countries with the largest percentage of electricity production from coal in 2013. *Data source*: World Bank Indicators http://data.worldbank.org/indicator/EG.ELC.COAL.ZS

2013, 75% of China's electricity was generated from coal (Figure 12.2), compared with 4% in France where 76% of electricity generation is from nuclear power. Depending on the technologies and operating practices at individual plants, the interquartile ranges of relative carbon intensity are 877–1130 CO_2-eq/kWh for coal and 8–45 CO_2-eq/kWh for nuclear electricity generation (Moomaw et al., 2011). On average, electricity from coal creates 60 times more CO_2 per unit of power than nuclear, ~100 times more than wind turbines and 250 times more than hydropower. Heavy reliance on coal for cheap electricity is, therefore, a major reason for China's high emissions.

Sector Footprints

A breakdown of emissions by sector *within* countries reveals the extent to which economies are driven by more or less carbon-intensive activities. Electricity and heat production plus agriculture, forestry and other land uses account for nearly half (49%) of all global carbon emissions (Figure 12.3a). The emission profiles of Low Income countries (e.g. Congo) are dominated by agriculture and forestry with relatively small contributions from industry. Conversely, in Upper Mid (e.g. Malaysia) and High Income countries (e.g. Germany) the largest sources are electricity and heat production, followed by transport and industry (Edenhofer et al., 2014). In the UK (2014), energy supply and transport each accounted for 23% of emissions, followed by business and industry (21%), public and residential buildings (16%), agriculture and land use change (9%), and waste plus flue-gas management (7%). Hence, by world standards, the UK has relatively low emissions from the agricultural sector, but higher than average emissions from transport and buildings (Figure 12.3b). Consequently, raising energy efficiency standards in buildings, transport and industry is one of the top priorities highlighted by the UK Committee on Climate Change (2015a).

Sub-sector and End-use Footprints

Sector footprints may be divided by sub-sector sources or end-user category. For instance, carbon emissions from transport can be allocated to aviation, road, railways and shipping, with further disaggregation of navigation into fishing, commercial, leisure and military uses. [Note that emissions from international aviation and shipping fall outside national inventories yet contribute about 5% of global emissions, and are rising rapidly. See Gençsü and Hino (2015).] Some end-users have very large CFs. For example, in 2013 iron and steel production in the UK accounted for 10% of all business sector emissions or about 3% of the national CF. Whilst the decline of the UK steel industry certainly harmed communities, it has contributed to falling national emissions.

The UK water sector also contributes ~3% when taking into account energy used for water abstraction, treatment, distribution and waste-water handling (Rothausen and Conway, 2011). At the same time, UK energy production is very water intensive (accounting for 60% of all UK water abstractions) so there is growing interest in opportunities for saving both energy and water at the so-called water–energy nexus (see Chapter 18). For example, using less hot water in the home saves energy in terms of smaller volumes of water requiring raw and waste treatment, pumping and distribution, as well as reduced energy for the water heating itself. Improved water efficiency at the level of a power plant might involve switching from a water-cooled to an air-cooled system. In this case, there is a trade-off between reduced energy efficiency of the plant but lower water demand, which could be beneficial in drought-stressed regions (Pacsi et al., 2013).

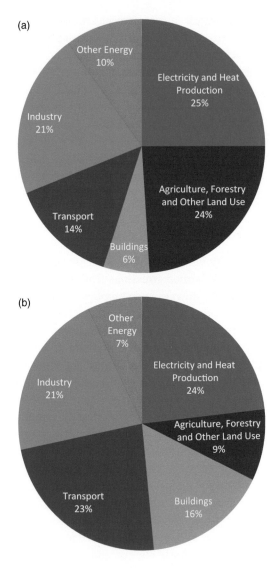

FIGURE 12.3 GHG emissions by major sector (a) globally (2010 data) and
(b) in the UK (2014 data). Data from IPCC (2014) and CCC (2015a).

Likewise, more stringent water quality standards for effluent dis-
charges to the environment may require more energy-intensive processes such as aeration and pumping. It has been estimated that
achieving the standards of the EU Water Framework Directive could
increase UK water sector CO_2 emissions by 110,000 tonnes per year
(Rothausen and Conway, 2011). Emissions could be even greater

when accounting for the carbon embodied in new construction or in chemicals used in treatment processes (Frijns, 2012). However, combined heat and power plants could be installed to harvest biogas from sludge digestion, thereby reducing both emissions and energy costs to the operator.

Household Footprints

The CF of a household reflects direct energy uses for space heating and motoring, as well as indirect emissions associated with purchased goods and services (see Druckman and Jackson, 2009). The household is a sensible unit for carbon accounting because many activities associated with the home have shared emissions (e.g. heating, car travel). Moreover, the household may be the focus for policies that are intended to reduce waste and improve efficiency. However, there can be significant variations in the carbon emissions of members *within* the same household. The number of occupants also matters: one person households tend to have higher per capita emissions than multiple-occupancy households. Those living in apartments have smaller CFs than same-sized families living in detached properties (Francis, 2004). Household energy use for space heating and air conditioning may also fluctuate between years due to the weather.

UK households contribute ~10% of direct national CO_2 emissions (CCC, 2015a). Measures aimed at cutting emissions include improving solid wall insulation, reducing waste and changing behaviours. For example, in 2012, there was an estimated 4.2 million tonnes of avoidable food waste (equivalent to 17 million tonnes of CO_2), and internal temperatures of homes have been rising since the 1990s. Another upward driver of emissions is the number of households, which is expected to grow from 27 million in 2014 to more than 31 million by 2030. Finance for more low-carbon power generation comes through renewable obligations added to consumer energy bills. Higher costs have to be weighed against genuine concerns about fuel poverty. In 2013, there were 4.5 million UK households spending at least 10% of their income on energy (CCC, 2015a).

Personal Footprints

The global average per capita CO_2 emissions is ~5 tonnes per year but there are enormous variations between and within nations. The top

TABLE 12.1 National per capita CO_2 emissions (tonnes per year) for the 12 highest and lowest emitters in 2011. *Data source*: World Bank

Highest emitters	Tonnes	Lowest emitters	Tonnes
Qatar	44.0	Burundi	0.021
Trinidad and Tobago	37.1	Chad	0.044
Kuwait	28.1	Congo, Dem. Rep.	0.050
Brunei Darussalam	24.4	Somalia	0.059
Aruba	23.9	Rwanda	0.063
Luxembourg	20.9	Central African Republic	0.063
United Arab Emirates	20.4	Malawi	0.079
Oman	20.2	Mali	0.080
Saudi Arabia	18.1	Ethiopia	0.084
Bahrain	17.9	Niger	0.084
United States	17.0	Eritrea	0.109
Australia	16.5	Uganda	0.111

emitting citizens live in Qatar (44 tonnes), Trinidad and Tobago (37.1 tonnes), and Kuwait (28.1 tonnes); the three populations with least emissions are Burundi (0.021 tonnes), Chad (0.044 tonnes) and Democratic Republic of Congo (0.050 tonnes) (Table 12.1). An individual's CF is shaped by a host of factors including age, socio-economic status, life-style, diet, and number of offspring. Over the course of a lifetime, recycling, switching to low-energy appliances, improving energy efficiency at home, driving less, and using a more fuel efficient car, might reduce the CO_2 footprint of a typical US citizen by ~490 tonnes. However, electing to have fewer children could save between 560 and 12,700 tonnes of CO_2 per child when taking into account the carbon legacy and average life expectancy of a child (Murtaugh and Schlax, 2009). In stark contrast, the same child born in Bangladesh in 2005 has a carbon legacy of 56 to 94 tonnes.

Personal CFs have a strong age dependency. In the UK, the 50–64 years age group has the highest levels of consumption of carbon-intensive goods and services (i.e. consumables, travel and food and drink, home and energy sources) (Royal Commission on

Environmental Pollution, 2011). In 2006, this group contributed 22% more CO_2 per capita than typical under 30 year olds. Assuming that successive generations have the same economic resources and life-style aspirations, an aging population could drive the national CO_2 footprint upwards. Even so, individual life-style choices make a big difference. For instance, a UK vegetarian who habitually cycles instead of drives, and who forgoes an annual holiday in California, would save nearly 7 tonnes of carbon per year compared with a meat-eating, car-driving, transatlantic flyer. Tim Peake (UK astronaut) has possibly the most carbon-intensive profession. One estimate suggests that the CO_2 footprint of rocket fuel production and burn each launch could be in the region of 630 tonnes.[1] This is equivalent to the combined annual CF of about 100 fellow citizens!

Some Indirect Ways of Assigning Carbon

Whether an astronaut or a zoologist, personal carbon accounting has to reflect both direct and indirect energy use. Direct carbon emissions include those from heating, lighting and transport. Indirect emissions are assessed using the full life-cycle cost of an activity or product even when originating from regions far from the consumer. Davis and Caldeira (2010) show, using a consumption-based accounting of CO_2 emissions, that Austria, France, Sweden, Switzerland and the UK have out-sourced more than 30% of their national emissions to imported goods. Conversely, about a quarter of all CO_2 emissions in China are linked to exports (Liu, 2015). Figure 12.4 shows the amount of embodied carbon flowing between nations as international trade. Over-all, the largest flows are from China into the United States and EU.

Carbon footprints are estimated for individual products using life-cycle assessment (Figure 12.5). For example, one short-sleeve T-shirt from China embodies about 6 kg of CO_2 and takes 1770 litres of water to produce (Zhang et al., 2015). The full life-cycle cost up to the point of purchase includes energy used for cotton cultivation, trans-portation of raw materials, spinning, knitting, dyeing, assembly and distribution of the finished product. Depending on how the T-shirt is washed and dried by the user throughout its lifetime, a further 0.04 kg (manual washing, no electric drying or ironing) to 8.62 kg (machine washing, electric drying and ironing) of CO_2 could be

[1] www.treehugger.com/renewable-energy/what-is-the-carbon-footprint-of-the-space-program.html [note conversion from imperial to metric tonnes]

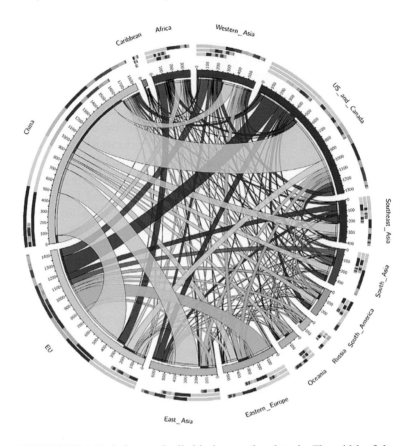

FIGURE 12.4 Emissions embodied in international trade. The width of the flow represents the emissions embodied in trade. The colour denotes the original production region, for example, the red flow represents the embodied emission produced by Africa and exported and consumed by other regions. *Source*: Figure 8 from Liu, Z. (2015). China's Carbon Emissions Report 2015. Belfer Center for Science and International Affairs. Harvard Kennedy School, Cambridge, MA.

emitted (Zhang et al., 2015). In this case, the carbon cost during the use phase (assigned to the T-shirt owner's national inventory) could be larger than that of the production and distribution phases. Yet more energy is used in the recycling or disposal phase.

The Carbon Trust (2006) provided an early assessment of carbon emissions tied to the provision of goods and services consumed in the UK (totalling 165.4 MtC each year). Rather than splitting emissions by sector, the report aggregated data from multiple supply chains. This revealed the goods and services with the largest

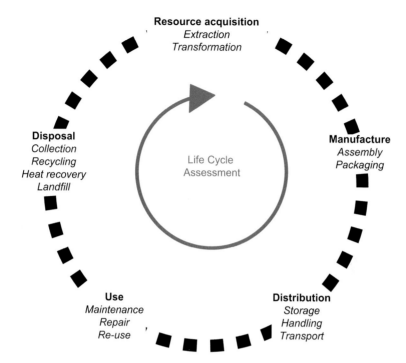

Resource acquisition
Extraction
Transformation

Disposal
Collection
Recycling
Heat recovery
Landfill

Life Cycle
Assessment

Manufacture
Assembly
Packaging

Use
Maintenance
Repair
Re-use

Distribution
Storage
Handling
Transport

FIGURE 12.5 Product life-cycle stages from cradle to grave.

emissions. In descending rank order they were recreation and leisure (31.6 MtC), space heating (24.0 MtC), food and catering (22.4 MtC), household (22.2 MtC), health and hygiene (21.7 MtC), clothing and footwear (16.1 MtC), commuting (13.1 MtC), education (7.9 MtC), other government functions (4.8 MtC) and communication (1.6 MtC). The analysis showed that nearly half of the emissions associated with recreation and leisure were due to transport.

City Footprints

Wolman's (1965) seminal work *The Metabolism of Cities* reflected growing concerns about the adequacy of water supply, disposal of sewage and the control of air pollution at the city scale. The metabolism concept has subsequently been extended to cover the larger ecological (including carbon) footprint of major cities. For example, Moore et al. (2013) estimate that the cropland and carbon sink (forest area) needed to offset Vancouver's ecological footprint is more than 10 million hectares, or 36 times the metro area itself. With annual per capita emissions of 10.9 tCO_2-eq, the CF of Vancouver is similar to that of Beijing (10.7 tCO_2-eq), Greater London (9.6 tCO_2-eq) and New York (10.5 tCO_2-eq) but more than Barcelona (4.2 tCO_2-eq) and Geneva (7.8 tCO_2-eq) (Kennedy et al., 2009; Sugar et al., 2012). Even larger CFs have been reported for Denver (21.5 tCO_2-eq) and the City of London (15.5 tCO_2-eq) (Kennedy et al., 2009; Minx et al., 2013). Such comparisons must be regarded with caution because of different years and methods of analysis. Nonetheless, there is evidence that the per capita CF of city dwellers depends on levels of car ownership, income, house/ household size, density of urban development, and availability of public transport. Hence, the low-density, mixed industrial/residential sprawl of Denver favours large emissions from ground transportation and heating, compared with the compact urban form and Mediterranean climate of Barcelona (Kennedy et al., 2009).

Returning to cricket, whatever the method of dismissal, the fielders must appeal to the umpire with 'how's that?' for the batsman to be given out. Likewise, concerted action will be needed to reduce CFs, whether at the T-shirt, personal or city scale. The next chapter shows the various policy and planning instruments being deployed as societies strive for the collective outcome of low-carbon development.

12.2 GROUP EXERCISE: CARBON LIFE-CYCLE ASSESSMENT

NOTES

ISO/TS 14067 specifies principles, requirements and guidelines for the quantification and communication of the CF of a product. It is based on International Standards on life-cycle assessment (ISO 14040 and ISO 14044) for quantification, and on environmental labels and declarations (ISO 14020, ISO 14024 and ISO 14025) for communication (International Organization for Standardization, 2013).

This activity examines the CF of a familiar commodity (bananas) using the ISO/TS 14067 cradle-to-retail-to-grave methodology. The task enables the identification of the most impactful life-cycle stages and hence possibilities for reducing emissions. The analysis also helps to pin-point those processes for which better data could reduce overall uncertainty in the CF. Finally, the task underlines the importance of clearly defining the system boundaries in order to be explicit about what is and is not included in the assessment.

Worksheet for Carbon Life-Cycle Assessment

1. Refer to Figure 12.5 and enter the stage for each process in a cradle-to-retail (Costa Rica to Norway) carbon assessment for 1 kg of bananas (Svanes and Aronsson, 2013).

Stage	Process	CO_2-eq (kg)
	Primary production	0.22
	Primary production infrastructure	0.0023
	Processing	0.017
	Packaging	0.080
	Transport to harbour	0.023
	Harbour storage and handling, Moin	0.016
	Overseas transport, pallet ships	0.75
	Harbour handling, Hamburg	0.017
	Transport from Hamburg to Kiel, truck	0.021
	Transport from Kiel to Oslo, passenger/cargo ship	0.059
	Transport from Oslo harbour to Ulven ripening plant	0.004
	Ripening at Ulven plant	0.008
	Banana waste until wholesale, production	0.012
	Transport to wholesale	0.006
	Storage in wholesale	0.007
	Transport to retail	0.008
	Retail in Norway	0.038
	Plastic bag used by some consumers	0.020
	Banana retail waste, production	0.038
	Cardboard box, retail waste	0.011
	Banana retail waste	0.004
	Plastic waste	0.002
	Total	**1.37**

2. What additional stages are needed to take the bananas to the 'grave'?

3. According to ISO 14067 the results of a cradle-to-retail assessment cannot be communicated to the public. Explain the reasoning.

4. What are the most uncertain entries in the cradle-to-grave carbon inventory?

5. Rank the life-cycle stages shown in the above table in terms of the size of their carbon equivalent (CO_2-eq) footprint.

6. Based on [5], identify possibilities for reducing emissions in the banana life-cycle.

LEARNING OUTCOMES

- To understand that CFs can be constructed for various units of assessment including individuals, cities, sectors and nation states.
- To recognise the role of carbon inventories and accounting in measuring progress and compliance with emission reduction targets.
- To be able to explain the key concepts of direct and indirect emissions, life-cycle assessment, carbon legacy and embodied carbon.

12.3 FURTHER QUESTIONS TO RESEARCH AND DISCUSS

Describe the livelihood and life-style of an adult with very low CF in (i) Chad and (ii) Germany.

What data would be needed to estimate the direct and indirect CF of the space industry (i.e. commercial, military and tourist)?

Describe the life-cycle stages of a smart phone, beginning with extraction of raw materials (including rare metals) for components and ending with handset disposal. Identify the most carbon intensive stage(s) in this bounded process.

12.4 FURTHER READING

Gurney, K.R. 2014. Recent research quantifying anthropogenic CO_2 emissions at the street scale within the urban domain. *Carbon Management*, **5**, 309–320.

Hertwich, E.G. and Peters, G.P. 2009. Carbon footprint of nations: a global, trade-linked analysis. *Environmental Science and Technology*, **43**, 6414–6420.

Scott, D., Gossling, S. and Hall, C.M. 2012. International tourism and climate change. *Wiley Interdisciplinary Reviews: Climate Change*, **3**, 213–232.

12.5 OTHER RESOURCES

BP Statistical Review of World Energy www.bp.com/en/global/corporate/energy-economics/statistical-review-of-world-energy.html [accessed 15/07/16]

UK Emissions by Sector: Committee on Climate Change www.theccc.org.uk/charts-data/ukemissions-by-sector/ [accessed 15/07/16]

UNFCCC National greenhouse gas inventory data for the period 1990–2013 http://unfccc.int/documentation/documents/advanced_search/items/6911.php?priref=600008730#beg [accessed 15/07/16]

World Bank Indicators http://data.worldbank.org/indicator [accessed 15/07/16]

WWF Carbon Footprint Calculator http://footprint.wwf.org.uk/ [accessed 15/07/16]

13

How to Decarbonise Economies?

TOPIC SUMMARY

The international community accepts that urgent action is needed to limit global warming to less than 2 °C. However, it is unclear how the requisite cuts in GHG emissions might be achieved. This is because there are many decarbonisation pathways involving different mixtures of financial instruments, technological innovation and behavioural change. The task is hampered by inexact knowledge about how the climate system might respond to emission reductions. Uncertainty also surrounds the cost of carbon, economic discount rate and feasibility of deploying unproven technologies (such as carbon capture and storage). Nonetheless, there are opportunities for improved energy efficiency in existing industrial-city complexes, and prototype eco-cities offer glimpses into how low-carbon lifestyles and urban planning might come together in the future.

BACKGROUND READING

Nakhooda et al. (2013) provide a summary of global finance to support developing countries mitigate and adapt to climate change. Warren et al. (2008) show how integrated assessment models are used to link the global economy, climate system behaviour, avoided damages to human and natural systems, and mitigation costs.

13.1 VISIONS AND PLANS FOR LOW-CARBON LIVING
Journey Starting Point and End

A plan of the new Chinese metropolis, Dongtan, was presented at the Shanghai World Expo in 2010 as a vision for the eco-friendly, zero-carbon city of the future. This was to be achieved through ambitious plans for recycling waste and by high energy efficiency for buildings and transport. The 500,000 city residents would be expected to play their part by relinquishing cars in favour of walking, cycling and public transport (unless driving cars powered by hydrogen or electricity) (Normile, 2008). Unfortunately, Dongtan has not taken shape as envisaged by planners and architects. There were concerns about the development's impact on neighbouring wetlands and rare birds but the project stalled ultimately because of lack of political support. Conventional high-rise apartment blocks have sprung up around the site, none of which are powered by renewable energy or built to high specifications. Journalist Malcolm Moore claims that Dongtan and other eco-city prototypes (such as Masdar City in Abu Dhabi) are *"green follies crippled by a central paradox: the more they enforce bothersome environmental rules, the less people want to live in them"*.[1] Others are more circumspect, pointing to the raised awareness of renewable energies, and push for greener building codes in the wider region (Reiche, 2010).

These far-sighted city projects provide a metaphor for the much greater challenge of decarbonising whole economies. The start and end-point for achieving this are broadly accepted: a move from ~10 GtC/yr to zero or even negative emissions to avoid dangerous climate change (see Chapter 11). There is less agreement about the most economically sensible and fair routes to achieve this outcome. The following sections describe some of the (i) policies and plans, (ii) financial instruments, (iii) technical innovations and (iv) changes in human behaviour that are required to decarbonise societies. Even if these measures are successful, some level of societal adaptation will likely be needed to manage residual climate impacts that have already accrued from historical emissions (Chapter 14). Given that cities account for ~70% of global energy consumption and house over half the world's population, they are sensible focal points for integrated mitigation–adaptation planning.

---

[1] www.telegraph.co.uk/news/worldnews/asia/china/9151487/Chinese-move-to-their-eco-city-of-the-future.html

Destination Decarb

The 2 °C global warming target is central to negotiations about curbing GHG emissions. Although there are reservations about the scientific basis of this limit (Knutti et al., 2016), 2 °C is highly ambitious (some now say even unrealistic) when devolved into the social and technical scenarios needed to achieve the emission cuts. To achieve this goal, there are basically two decarbonisation pathways: (i) immediate slashing of GHG emissions ('pay now'); or (ii) slower reductions after huge investment in bio-energy and carbon capture and storage (CCS) technologies ('pay later') (Tollefson, 2015). The former envisages rapid replacement of energy supplies from fossil fuels by biomass and other renewables (Figure 13.1, left column); the latter sees fossil fuels peaking in the 2040s whilst expanded nuclear power, forests and CCS are established to achieve *negative* emissions by the 2070s (Figure 13.1, right column). In this case, CO_2 is actually withdrawn from the atmosphere. [Note that CCS is discussed alongside geoengineering technologies in Chapter 15.]

Integrated assessment models (IAMs) link modules of socio-economic development, technological change, climate responses and mitigation costs to simulate decarbonisation scenarios (e.g. Riahi et al., 2007; Warren et al., 2008). The IAMs also reveal where climate policy may be susceptible to gaps in scientific knowledge. For example, much uncertainty surrounds the climate sensitivity (β) of natural carbon sinks to rising temperatures (e.g. forests, ocean plankton and downwelling). This matters because carbon sequestration by terrestrial and ocean ecosystems is a 'free' good that reduces the costs of implementing abatement technologies – the higher the value of β the lower the need for, and cost of mitigation. If $\beta = 0.3$, natural sinks may sequester nearly all emissions, but if $\beta = 0$, technologies may be required to meet a shortfall totalling 696 $GtCO_2$ over the period 2050–2099 (Fuss et al., 2013). Even if negative emissions and a 2 °C stabilisation target can be achieved, the response of the oceans will lag behind the atmosphere. This means that sea level rise caused by the thermal expansion of the oceans could continue for several more centuries after peak emissions (Tokarska and Zickfeld, 2015).

Policies and Plans to Reduce Emissions

The 'pay now' for deep cuts strategy is preferable if the goal is to minimise the risk of irreversible climate change. This implies that the

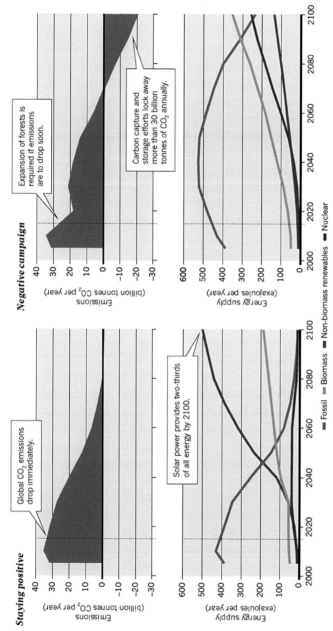

FIGURE 13.1 Global energy-supply pathways to achieve 2 °C via early (left) and deferred (right) emissions reductions.
Source: Tollefson (2015)

CO_2 concentration should peak below a target of 650 ppm to achieve stabilisation of radiative forcing by 2100. To achieve a more stringent target of 1.5 °C global warming by 2100 requires that the CO_2 concentration peaks at 450 ppm by 2060. These stabilisation concentrations equate to the Representative Concentration Pathways (RCPs) 4.5 and 2.6, respectively (Figure 6.3), but there are many different socio-economic trajectories to achieve these outcomes. Under business as usual, global emissions from fossil fuel combustion and land use changes most closely resemble RCP 8.5 (Fuss et al., 2014). The RCP 2.6 target now looks implausible given an annual mean growth rate of 2.1 ppm atmospheric CO_2 over the past decade and concentrations at Mauna Loa passing the 400 ppm mark in 2015 (Figure 2.5).

Given a global warming target of 2 °C at a stabilisation concentration of 450 ppm, the question becomes: what is the most cost-effective mitigation pathway(s) to achieve RCP 4.5? Thomson et al. (2011) analysed different transformations of the energy system, land use and global economy to attain this target. Each mitigation scenario was compared to a reference with the same population (8.7 billion people) and tripling of global primary energy consumption by 2100 but without GHG reduction policies. Some mitigation pathways do not invoke CCS or nuclear power, or assume that all countries will participate in mitigation. However, these require higher carbon prices and greater global productivity of bio-energy crops (with knock-on effects for the cost of food due to conversion of some crop and pasture lands). Ironically, there might also be substantial deforestation of land for bio-energy production. It turns out that the most cost-effective technological pathway to RCP 4.5 requires a mix of bio-energy with CCS (Figure 13.2). Even a global transition to more healthy, low-meat diets would contribute to smaller mitigation costs (Stehfest et al., 2009).

RCP4.5 cost-minimising stabilisation scenarios assume that all nations participate in emissions reduction and that there is one global price for carbon emissions. In practice, the present set of Intended Nationally Determined Contributions (INDCs) show major variations in mitigation strategies at the country scale (Chapter 11). For example, Brazil and China have strong reliance on sequestration by forests and expansion of renewable energy supplies, whereas the EU and United States are seeking efficiency gains from buildings and transport. Some of the assumed technological advances look questionable even now. For instance, in 2015, the UK government scrapped a £1 billion CCS fund, casting doubt over the future of this

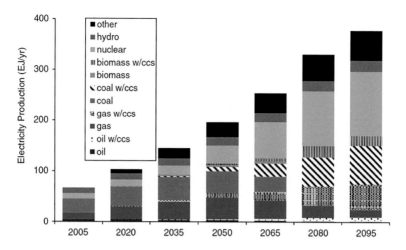

FIGURE 13.2 Electricity generation by technology type in the RCP4.5 scenario. Note the assumed growth of nuclear energy production and carbon capture and storage (CCS) for coal and gas electricity generation. *Source*: Thomson et al. (2011)

technology as a central plank in the nation's mitigation strategy (see Figure 13.3). A zero-carbon target for new houses has also fallen by the wayside and the UK Feed-in Tariffs scheme for renewables has been cut back. Conversely, the EU and California have set rather ambitious zero net energy targets for new buildings by 2020.

Climate Finance and Initiatives to Reduce Emissions

The *Stern Review* (Stern, 2007) made the economic case for sharp and immediate cuts in GHG emissions. The author reasoned that the damages caused by climate change would outweigh the costs of mitigation. Subsequent critical analysis challenged this central claim because the economic discount rates used for valuation (averaging ~1.4%) were lower than in previous studies (Nordhaus, 2007). Low rates magnify the weight attached to climate impacts in the distant future, thereby rationalising deep cuts in emissions now. Nonetheless, expectations that industrialised nations should lead mitigation efforts remain founded on the fact that they are most to blame for historical GHG emissions, yet incurring least burden of future climate damages (Srinivasan, 2010). At the 2015 UNFCCC talks in Paris, wealthy countries re-stated a pledge to commit US$100 billion annually by 2020 to tackle the causes and consequences of climate change.

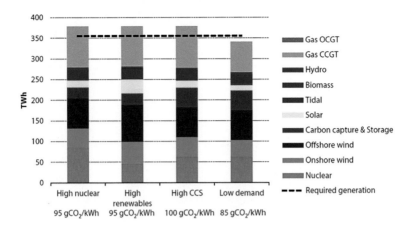

FIGURE 13.3 UK energy generation (TWh) by source for 2030 under four scenarios with central deployment rates for low-carbon technologies: (1) High nuclear (assumes three new nuclear power plants providing 24% generation compared with 21% in other scenarios); (2) High renewables (assumes more sites are found for onshore wind and solar to provide 61% generation compared with 45% in other scenarios); (3) High CCS (technology is deployed for 9 GW compared with 4 GW in other scenarios) – now looking unlikely given the cancellation of trial projects in 2015; and (4) Low demand (assumes greater uptake of efficiency measures). The figures below each scenario caption are the average grid intensity (gCO$_2$/kWh). Source and copyright: Committee on Climate Change (2015b)

There are three main routes by which climate finance flows from developed to developing countries to assist their mitigation and adaptation efforts (Table 13.1). Beginning with the largest, multilateral finance is disbursed via World Bank Climate Investment Funds to interventions such as the Pilot Program for Climate Resilience (PPCR), Clean Technology Fund (CTF) and the Forest Carbon Partnership Facility (FCPF). These are intended to assist low-carbon and climate resilient development pathways. Other finance for environmental projects is administered by the UN Global Environment Facility (GEF) or REDD programme. The UN Adaptation Fund is financed by a 2% levy on sales of emission credits through the Clean Development Mechanism (CDM) of the Kyoto Protocol. Next in size (yet still in the realm of US$ billions) are the bilateral funds administered mainly through development agencies on a country-to-country basis (e.g. Germany's International Climate Initiative, Norway's International Forest Climate Initiative and the UK's International Climate Fund). Finally, there are the national climate change initiatives,

TABLE 13.1 Multilateral, bilateral and national climate funds. Adapted from Nakhooda et al. (2013)

Acronym	Multilateral funds and initiatives
AF	Adaptation Fund (GEF acts as secretariat and WB as trustee)
CBFF	Congo Basin Forest Fund (hosted by AfDB)
CDM	Clean Development Mechanism (implemented under the Kyoto Protocol)
CIF	Climate Investment Funds (implemented through WB, ADB, AfDB, EBRD and IADB)
CTF	Clean Technology Fund (implemented through WB, ADB, AfDB, EBRD and IADB)
FCPF	Forest Carbon Partnership Facility
FIP	Forest Investment Program (implemented through WB, ADB, AfDB, EBRD and IADB)
GCCA	Global Climate Change Alliance
GCF	Green Climate Fund
GEF	Global Environment Facility
GEF 4	GEF Trust Fund Fourth Replenishment
GEF 5	GEF Trust Fund Fifth Replenishment
GEEREF	Global Energy Efficiency and Renewable Energy Fund (hosted by EIB)
JI	Joint Implementation (implemented under the Kyoto Protocol)
LDCF	Least Developed Countries Fund (hosted by the GEF)
PPCR	Pilot Program for Climate Resilience (implemented through WB, ADB, AfDB, EBRD and IADB)
SCCF	Special Climate Change Fund (hosted by the GEF)
SCF	Strategic Climate Fund (implemented through WB, ADB, AfDB, EBRD and IADB)
SREP	Scaling Up Renewable Energy Program (implemented through WB, ADB, AfDB, EBRD and IADB)
UNREDD	United Nations Collaborative Programme on Reducing Emissions from Deforestation and Forest Degradation

cont.

TABLE 13.1 (cont.)	
	Bilateral funds and initiatives
FSF	Fast Start Finance (Japan)
GCCI	Global Climate Change Initiative (United States)
ICF	International Climate Fund (UK)
ICFI	International Climate Forest Initiative (Norway)
ICI	International Climate Initiative (Germany)
IFCI	International Forest Carbon Initiative (Australia)
	National climate funds
AF	Amazon Fund
BCCTF	Bangladesh Climate Change Trust Fund
FONERWA	Rwanda National Climate and Environment Fund
GRIF	Guyana REDD+ Investment Fund
ICCTF	Indonesia Climate Change Trust Fund
MCCF	Mexico Climate Change Fund
PSF	Philippines People's Survival Fund

Key: ADB: Asian Development Bank; AfDB: African Development Bank; EBRD: European Bank for Reconstruction and Development; EIB: European Investment Bank; IADB: Inter-American Development Bank; WB: World Bank.

which may be backed by a mix of international finance, domestic budgets and private sector investment (e.g. Brazil's Amazon Fund or Mexico's Climate Change Fund).

Despite the scale and ambition of proposed climate funds, questions remain about the additionality of some finance (i.e. how much of the public sector money is just rebranding development budgets as climate finance). Proliferation of funding mechanisms also means that it is difficult to track flows and verify investment outcomes. Ideally, finance from different sources would be aligned to avoid duplication as well as to maximise social co-benefits, such as improved gender equality (Nakhooda et al., 2013). However, it is recognised that globally comprehensive and harmonised mitigation actions would achieve emission reductions at the lowest aggregate economic cost if investment flowed to where it is least expensive to achieve unit reductions in the carbon burden (Edenhofer et al., 2014). The price of carbon is a critical variable in determining those aggregate economic costs once cascaded to all sectors of the global economy.

Carbon Taxation

Taxing carbon emissions sends strong financial signals to invest in clean, low-carbon technologies. The EU Emissions Trading System (EU ETS) applies a 'cap and trade' principle to set a maximum level of GHG emissions from regulated industries in all EU member states plus Iceland, Norway and Lichtenstein. It was the first such major scheme of this kind and is now in Phase 3 of development (Figure 13.4). In 2013, the cap was fixed at 2080 million tonnes of CO_2, with allowances set to reduce by 38 million tonnes annually until 2020. The 'traded' sectors cover about 45% of the EU's GHG emissions, primarily from electricity generation, other energy-intensive industries and aviation. Companies within the cap have to buy or trade allowances depending on how much they emit. By 2020, emissions covered by the EU ETS are expected to be 21% lower than in 2005. Under Phase 2 of the EU ETS, nations were also able to buy carbon offsets through the CDM and Joint Implementation projects under the Kyoto Protocol – effectively boosting the cap by importing additional carbon credits.

According to the European Commission, the EU ETS yielded an 8% reduction in regulated GHG emissions over the period 2005–2010.

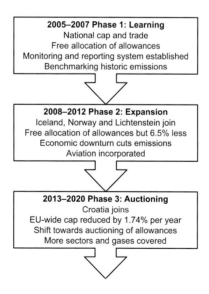

FIGURE 13.4 Development of the EU Emissions Trading System over the first three phases 2005–2020.

Opponents claim that the US$287 billion costs could have achieved much greater cuts had investment been targeted at the most polluting installations across Europe. During Phase 1 (2005–2007) over-allocation of allowances lead to windfall profits for some companies. In 2008 the carbon price was nearly €30/tCO$_2$; by January 2013 it was less than €3/tCO$_2$. This fall was partly explained by global recession and the ensuing reduction in emissions from energy-intensive sectors. Unfortunately, significant volatility in the price of carbon deters long-term investment in low-carbon technologies. Hence, in May 2014, the UK government set a carbon floor (minimum) price of £18.08 for 2015/16 to incentivise expansion of renewable energy and thereby contribute to carbon reduction targets.

Technical Innovation and Behaviour Change to Reduce Emissions

To lower emissions, there must also be innovation in both energy supply and demand-side low-carbon technologies alongside market mechanisms (van Vuuren et al., 2016). Supply-related technologies currently at the demonstration phase include CCS and hydrogen fuel vehicles; others such as nuclear power, heat pumps, offshore wind and electric vehicles are considered commercially proven and at the deployment stage (CCC, 2015b). Basic research and feasibility studies are underway for micro-generation and distributed energy sources such as heat extraction from drinking water (Blokker et al., 2013) or rooftop photovoltaic systems for houses and schools (Koo et al., 2014).

Given that the majority of new construction over coming decades will be in the megacities of the developing world, there are opportunities for designing new power grids, transport systems and homes with carbon saving in mind (Chapter 10). For instance, Salazar and Meil (2009) estimate that the carbon life-cycle balance of a typical wood-frame home (in Ottawa) is 5.2 tons of CO$_2$-eq permanently removed from the atmosphere (i.e. a net carbon sink) compared with 63.4 tons CO$_2$-eq sourced from a conventional brick-clad house. Others point to the huge energy savings that could be achieved from existing buildings. For example, it is estimated that efficiency measures could achieve a 60% reduction in emissions from buildings in New York City (Wright et al., 2014). Meanwhile, gardens and parklands in cities can be used as carbon sinks to offset emissions from households and transport (Zhao et al., 2011). One study found that carbon neutrality is achieved when the 'natural fraction' of a city exceeds about 80% of the metro-area (Nordbo et al., 2012).

'Lock-in' by existing energy, waste and transport infrastructure hinders the diffusion of new carbon-saving technologies. This is because institutions, regulatory systems and fossil-fuel-based technologies have co-evolved in ways that now favour centralised electricity generators rather than distributed, micro-generation by renewables or combined heat and power (Unruh, 2000). Other technologies such as hydrogen-based systems would require wholescale changes in infrastructure to support distribution and allay safety concerns. More generally, the sector can be regarded as a social-technical system that connects personal behaviours with new technologies and patterns of energy consumption. For example, the energy saving from an ultra-low emission vehicle will only achieve intended carbon reductions if the total distance driven does not rise to cancel efficiency gains. Likewise, a shift to greater homeworking may save emissions from avoided commuting on one hand but increase home energy consumption on the other. In the UK, the tipping point to achieve net carbon saving is when a commuter works at home rather than travelling further than 7 km by car or 25 km by train to work (Carbon Trust, 2014).

Concluding Remarks

Three elements are needed to decarbonise an economy, whether at the global or national scale: (i) an agreed target for GHG emission reductions that is intended to avoid dangerous climate (Chapter 11); (ii) techniques for measuring carbon footprints and thereby identifying 'hotspots' that warrant most urgent attention (Chapter 12); and (iii) economic policies that incentivise investment in low-carbon energy sources, energy efficiency gains, and less carbon-intensive behaviours (this chapter). Even if the most optimistic scenarios for CO_2 reductions are realised, some climate change impacts are likely due to the legacy of past emissions. Chapter 14 explains the steps that can be taken to adapt to this unavoidable climate change.

13.2 GROUP EXERCISE: CARBON MANAGEMENT PLAN

NOTES

This activity works through the stages of a carbon management plan. The target business could be a well-known product brand, service provider or commodity supply chain. This can be customised according to the group. Students may be curious about the most energy intensive activities and opportunities for carbon reductions on their own campus (see, for example, the Loughborough University carbon management plan under 'Other Resources'). Many universities have now developed carbon plans (see also the *EcoCampus* scheme and Amaral et al., 2015). Whatever the chosen business, the exercise is most effective when students refer to actual data on energy consumption, emissions and costs. The worksheet can be completed step-by-step with group discussion, or as self-directed study.

Worksheet for Carbon Management Plan

Review the assigned carbon management plan then complete the following fields.

1. List the main elements of a carbon management plan.

2. List the main sources of emissions that are directly allocated (Tier 1), due to purchased energy (Tier 2), or indirectly incurred (Tier 3) by the business.

   Tier 1:

   Tier 2:

   Tier 3:

3. List the most carbon intensive activities and/or facilities operated by the business.

4. Give examples of data that could be used to quantify Tier 1 to 3 emissions.

5. What factors might cause year-to-year variations in emissions by the business?

6. Suggest ways in which the chosen business could reduce emissions.

7. Suggest ways in which the carbon management plan might be financed.

8. What indirect benefits might accrue from a business carbon management plan?

LEARNING OUTCOMES

- To appreciate that decarbonisation pathways are shaped by many factors including assumptions about the climate sensitivity to emissions, available technologies to increase low-carbon energy supplies and/or reduce demand, policies and economic instruments to achieve intended targets.
- To be aware of the landscape of global finance that supports low-carbon development in emerging economies, as well as the key features of the EU Emissions Trading System.
- To recognise that the benefit of new technology for carbon emissions is strongly determined by social and behavioural relationships with those technologies.

13.3 FURTHER QUESTIONS TO RESEARCH AND DISCUSS

How might the vision of a zero-carbon city be realised?
By how much and by when should greenhouse gas emissions be reduced?

Compare the merits of reducing carbon emissions via improved energy efficiency against increasing the supply of low-carbon energy through technological innovation.

13.4 FURTHER READING

Barrett, S. and Dannenberg, A. 2012. Climate negotiations under scientific uncertainty. *Proceedings of the National Academy of Sciences*, **109**, 17372–17376.

Premalatha, M., Tauseef, S.M. and Abbasi, T. 2013. The promise and the performance of the world's first two zero carbon eco-cities. *Renewable and Sustainable Energy Reviews*, **25**, 660–669.

Tol, S.J. and Yohe, G.W. 2006. A review of the *Stern Review*. *World Economics*, **7**, 233–250.

13.5 OTHER RESOURCES

Climate Funds Update www.climatefundsupdate.org/ [accessed 15/07/16]

Deep de-carbonisation of electricity grids http://judithcurry.com/2015/11/29/deep-de-carbonisation-of-electricity-grids/ [accessed 15/07/16]

EcoCampus Scheme www.eauc.org.uk/ecocampus [accessed 15/07/16]

Loughborough University Carbon Management Plan (2010) www.lboro.ac.uk/media/wwwlboroacuk/content/sustainability/downloads/2.3.4%20Carbon%20Managment%20Plan.pdf [accessed 15/07/16]

How Is It Possible to Adapt to an Uncertain Climate?

TOPIC SUMMARY

The Intergovernmental Panel on Climate Change (IPCC) defined adaptation as *"the process of adjustment to actual or expected climate and its effects"*. This is fine in theory, but what 'adjustments' can be made in practice when the 'expected climate and its effects' are so vague? By placing decision-making (rather than climate scenarios) at the heart of the process, emphasis is shifted onto understanding the concerns of stakeholders, vulnerability of the system, and available options. From this perspective there can be measures and strategies that deliver benefits whatever the climate outlook. The scenario then becomes a means of testing how different options perform under climate change (alongside other non-climatic pressures). Options can be arranged as adaptation pathways that are followed in response to emerging climate risks. Scientific experiments can also be designed to translate high-level adaptation objectives into specific guidance for practitioners.

BACKGROUND READING

Adger et al. (2009) consider the social limits to adaptation and explain how these are, amongst other factors, ultimately dependent on the goals of the adjustment process. Hallegatte (2009) offers a very practical set of strategies for adapting to climate change under uncertainty. Brown and Wilby (2012) explain how climate scenarios can help discover circumstances under which an adaptation might fail.

14.1 ROBUST DECISION-MAKING
Battling the Uncertainty Monster

A character in Benjamin Disraeli's *The Wondrous Tale of Alroy* (1833) proclaims "*I am prepared for the worst, but hope for the best*". The novel is based on a thirteenth-century uprising by the Jewish community against Seljuk Sultan Muktafi. The words were declared by one of the insurgents about his readiness for battle. More recently, the quote has become a rallying cry for the 'war' on climate change as nationally pledged emissions reductions are unlikely to keep global warming below 2 °C (Averchenkova and Bassi, 2016) (Chapter 13). One analysis claims that this threshold could be crossed as early as 2036 (Mann, 2014); others hope that, after overshooting the 2 °C target, nations will adapt, then recover in coming centuries (Parry, 2009). Meanwhile, the struggle is to adjust *proactively* in ways that minimise expected harms and maximise benefits to society *without* certainty about the future climate conditions. Thanks to the 2015 Paris Conference of the Parties, the annual 'war chest' for adaptation (and mitigation) could be US$100 billion by 2020.

Chapter 6 explained the limitations of conventional predict-then-act approaches to climate risk management. By the end of the scenario and modelling chain, the span of plausible regional climate change (e.g. Figure 6.5) and associated impacts (e.g. Figure 6.7) can be so wide that it bewilders decision-makers (Figure 14.1). Curry and Webster (2011) observe that this climate uncertainty 'monster' may invoke denial about the need to act, instil doubt about ability to detect changes, or cause delay until more scientific evidence is gathered. Claims that the uncertainty can be reduced through more research are not supported by the history of climate science, which shows that increased model complexity has not generally translated into narrower uncertainty (Chapter 4). The most prudent option is to assimilate the uncertainty – to embrace it as a given, and to develop adaptation strategies that are good enough for most of the time, but not necessarily optimal. Robust decision-making (RDM) or 'satisficing' frameworks have emerged as promising ways to achieve just that (Lempert and Collins, 2007).

This chapter describes different approaches to managing climate risk and explains where they sit within a RDM framework. The resulting adaptation strategies are surprisingly pragmatic, yet responsive to evolving climate conditions and societal needs. Three case studies demonstrate how anticipatory actions can be tested and

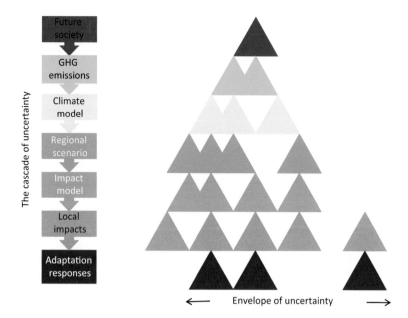

FIGURE 14.1 The cascade of uncertainty proceeds from different socio-economic and demographic pathways, their translation into concentrations of atmospheric greenhouse gas (GHG) concentrations, expressed climate outcomes in global and regional models, translation into local impacts on human and natural systems, and implied adaptation responses (see an alternative visualisation in Figure 6.5). The number of triangles at each level symbolises the growing number of permutations and hence expanding envelope of uncertainty. Missing triangles represent incomplete knowledge or sampling of uncertainty. Note that the range of uncertainty is most likely underestimated. Adapted from Wilby and Dessai (2010).

deployed despite uncertainty about climate change. The examples cover water resource, flood risk and conservation management. The final section outlines ways of discerning whether adaptation might be successful.

Adaptation Frameworks and Options

Numerous conceptual frameworks have been devised to help decision-makers work through the steps needed to adapt to climate change. The risk and uncertainty framework of Willows and Connell (2003) was introduced in Figure 9.3, and is amongst the most widely used. Having set out the objectives and decision-making criteria as well as assessed key risks, the next task is to identify and appraise adaptation options. There are many ways in which this can be done, but the most common approach is to undertake stakeholder consultation and vulnerability assessment (Figure 14.2). Such participatory approaches provide local knowledge of the climatic and non-climatic pressures at work, as well as the various adaptation options that might be effective given existing socio-cultural, institutional and technical barriers. For example, one study of agriculture-dependent households in northeast Ghana found (as with many others) that to obtain improved varieties of crops communities first needed access to micro-credit schemes (Antwi-Agyei et al., 2014).

Table 14.1 lists various adaptation domains. These generic activities are not mutually exclusive and may be applied individually or as a portfolio. For example, increasing resilience of public water supplies to future drought might involve construction of new or enlarged reservoirs (infrastructure), combined with water saving appliances in the home (retrofit), targets to reduce leakage in the distribution network (institutional), higher tariffs for non-essential water consumption (financial), and public awareness campaigns to curtail domestic water use (behavioural). RDM strategies deploy measures that perform reasonably well (i.e. minimise regret) across a range of plausible futures (Table 14.2). Here, regret is defined as the difference between the present benefit of a strategy in *any* state of the world and the expected benefit of the optimum strategy in *each* state of the world (Lempert and Collins, 2007). 'Low-regret' strategies are often advocated because they deliver benefits whatever the climate scenario – such as an accurate drought forecasting system that would be just as beneficial in the 2050s as now. Other strategies such as 'safety margins' try to incorporate an allowance for climate change, but the size of that margin depends on a host of factors, not least cost and tolerance to risk (e.g. Environment Agency, 2011).

The eventual set of options has to support agreed adaptation objective(s) (Figure 14.2). The intent may be to avoid loss of human life (from flooding or heatwaves), to reduce economic damages (from droughts or coastal erosion), or loss of biodiversity (as a result of invasive species or habitat degradation). Mathematical models can be used to test the performance of options under prescribed future climate and non-climatic pressures (e.g. Whitehead et al., 2006). For instance, hydrological modelling shows that the most effective strategy for minimising future restrictions on public water supply in East Devon, UK, could be a mixture of increased reservoir storage

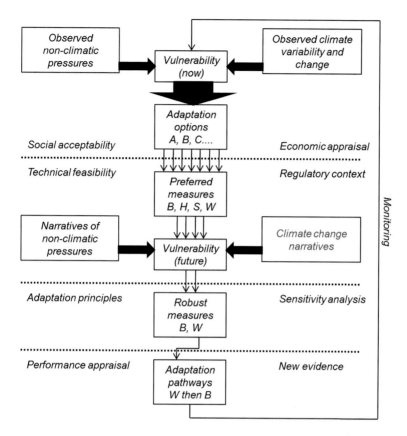

FIGURE 14.2 A conceptual framework for the assessment of robust adaptation options. The process begins by reviewing vulnerabilities alongside a suite of counter measures (labelled A, B, C...). Options are screened against overall adaptation objectives and acceptability (in terms of social, economic, etc. criteria). The sub-set of preferred measures (labelled B, H, S, W) are next subjected to stringent testing to determine how well they perform under plausible narratives of climatic and non-climatic pressures. These sensitivity (sometimes called 'stress') tests are typically based on expert opinion, model experiments, or a combination of both. The most robust measures (labelled B, W) are then arranged as an adaptation pathway (labelled W then B), which includes sequencing and trigger points for their activation. For instance, W might be triggered first at sea level X, followed by B at sea level Y. Long-term monitoring of adaptation outcomes is needed to check that measures are sufficient for the unfolding rate and types of change. Further adjustments are made in the light of new evidence. *Source*: Wilby and Dessai (2010)

TABLE 14.1 Adaptation domains and example actions. Modified from Wilby et al. (2009)

Domains	Example adaptation activities
Infrastructure	Build defences against coastal erosion and flood risk from sea level rise; enlarge reservoirs to store more water to counter variable rainfall; design buildings to cope with extreme weather
Natural resource management	Review natural resource availability within long-term (water-energy-food) plans; adjust resource allocations or operating rules to reflect changing conditions; reduce non-climatic pressures on natural systems (such as pollution); promote efficiency, re-use and recycling
Retrofit	Upgrade existing infrastructure to cope with more frequent and/or severe extreme events (such as higher capacity drainage systems); incorporate more shade, green spaces and water features in cities to provide refuges during heatwaves
Behavioural	Use forecasts to improve preparedness and guide risk reduction strategies (for heatwaves, floods and droughts); adopt resource conserving behaviours (see *Financial* measures below)
Institutional	Adjust performance standards for service providers (such as for water and energy reliability); diversify and/or strengthen supply chains; formulate and monitor adaptation plans for essential services (such as water, energy, transport and communications)
Sectoral	Provide advice for practitioners (such as climate safety margins for engineers); upgrade codes to reflect changing climate (such as building standards); promote knowledge sharing and cooperation within and between sectors

cont.

TABLE 14.1 (cont.)	
Domains	Example adaptation activities
Communication	Raise awareness of climate risks and opportunities through the media, professional and public bodies; disseminate information using formats and channels that are meaningful to diverse audiences
Financial	Incentivise climate smart behaviour through pricing signals (such as water tariffs); spread climate risk by insurance services; protect vulnerable livelihoods through forecast-based finance schemes and pay-outs; diversify the livelihoods of low-income households

with reduced per capita water demand (Lopez et al., 2009). Water supply and demand are balanced, so this combination is robust to a wide range of climate model scenarios. However, real systems are more complex because they trade off multiple objectives under climate change – such as meeting the water needs of people with the needs of the environment (see Figure 14.3 below, or Poff et al., 2016).

Case 1: Stress Testing an Adaptation Measure for Denver Water

RDM frameworks help to explore the important questions: How does my system work and when might my policies fail? (Weaver et al., 2013). In the case of Denver Water, Colorado, these concerns were addressed through a four-step collaborative decision-support process (Yates et al., 2015). These involved (i) identifying the management practice(s) to be evaluated; (ii) modelling the water supply through physical representation of the hydrological cycle; (iii) simulating the operation of the water collection and distribution systems in the context of legal water rights; and (iv) stress testing the system using narratives of future climatic and non-climatic conditions. Their purpose was to explore the performance of the adaptation option(s) in supporting overall water supply to the city of Denver.

Figure 14.3a gives a sense of the complex mix of pressures and demands on the flows of the Upper Colorado River Basin. Step 1 of the adaptation appraisal focused on one drought response measure, the

TABLE 14.2 Robust strategies for adapting to uncertain climate change. After Hallegatte (2009)

Strategies	Definition	Examples
Low regret	Measures that yield benefits regardless of the climate outlook	Real-time monitoring and forecasting of flood hazards; upgrade infrastructure to higher specification on replacement
Reversible	Measures that keep as low as possible the cost of being wrong	Easy to retrofit coastal defences that enable cheap upgrades if sea level rise accelerates; restrictive urban planning
Safety margin	Extra 'headroom' to absorb climate change and reduce vulnerability at least cost	Precautionary allowance applied to peak river flows or extreme sea levels to account for climate change when designing infrastructure
Soft	Institutional or financial measures that plan for and/or spread risks	Land use zoning, early warning systems and insurance schemes; land set aside for floodwater retention
Shorten time horizon	Counter uncertainty in future climate conditions by reducing the lifetime of investments	Cheap or modular infrastructure that can be replaced on shorter cycles as risk of flooding increases
Integrated	Manage positive and negative side-effects of adaptation, including trade-offs with mitigation or benefits across different sectors	Restore salt marsh or mangrove swamps to enhance biodiversity whilst improving coastal flood/erosion protection

(a)

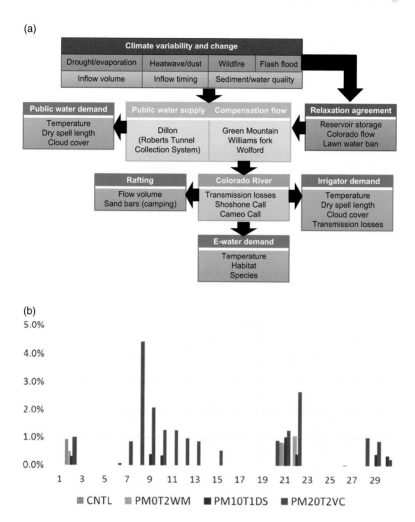

(b)

FIGURE 14.3 (a) Multiple pressures and management objectives affecting flows in the Upper Colorado River Basin. (b) Percentage increases in storage within the Upper Colorado basin when invoking the Shoshone Call Relaxation Agreement under control (CNTL), modest warming (PM0T2WM), dust on snow (PM10T1DS) and vegetation change (PM20T2VC) narratives. *Source*: Yates et al. (2015)

Shoshone Call Relaxation Agreement (SCRA). Under the terms of the SCRA, Denver Water is able to secure additional water from Excel energy when faced by certain drought conditions. Step 2 represented the behaviour of the system using the Water Evaluation and Planning (WEAP) tool. The model includes snowmelt and runoff, reservoir storage and water transfers over the Continental Divide to Denver. Step

3 translated the various water rights and terms of the SCRA into 'if then' rules within the WEAP model. Finally, step 4 assessed the effect of the SCRA on system-wide water storage, compared to a baseline with no SCRA. This evaluation was performed for the present climate and three narratives of future climate and catchment conditions.

Contrary to expectations, there were only modest gains in water stores due to the SCRA when compared with the control situation (CNTL in Figure 14.3b). Under present climate conditions the SCRA might be triggered twice in 30 years, each time increasing storage by ~1%. Under the PM0T2WM narrative of mild warming (+2 °C) and no change in precipitation, the SCRA had roughly the same impact. With less warming (+1 °C) but a 10% reduction in precipitation, combined with dust-on-snow events (contributing to earlier, more rapid snowmelt) in PM10T1DS, the SCRA activated during 7 out of 30 years but delivered <1% more storage. Under mild warming (+2 °C) and 20% less precipitation with 5% forest die-back in PM20T2VC, the SCRA was invoked in 16 out of 30 years and yielded less than 2% additional water storage. Based on these findings, the study concluded that the specific flow thresholds and conditions for triggering the SCRA make the agreement quite weak as a drought management tool.

Case 2: Adaptation Pathways for the Thames Estuary

The Thames Estuary 2100 Plan considered multiple options for defending London against future flooding by sea level rise, storm surge, river and urban runoff. The plan has three main stages: (i) maintain or upgrade existing flood defences whilst safeguarding spaces for future floodwater storage (2010–2034); (ii) renew and replace existing tidal defences (2035–2070); and (iii) maintain the existing system and construct a new barrier (2070 onwards). Given the uncertainty in the rate of sea level rise and possible changes in storm surge risk, these high-level options can be brought forward or delayed as required.

The whole process is envisaged as a set of option pathways, where decisions are triggered by indicators such as a threshold sea level (Figure 14.4). For example, one pathway (shown with a blue line) involves raising existing defences up- and downstream, creating flood storage, and over-rotating the Thames Barrier (Figure 14.5). This pathway could accommodate sea level rise of about 2 m. Beyond that level, more radical options (such as a second barrier in the river or barrage in the estuary) would be required. Moreover, a lead time

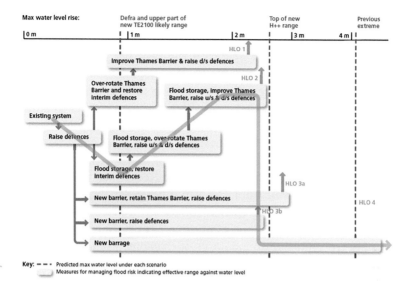

FIGURE 14.4 High-level options (HLOs) and adaptation pathways developed by the Thames Estuary 2100 strategy. The horizontal axis shows thresholds (of sea level rise) that trigger progressively more radical measures. H++ is a plausible, high-end climate change scenario that is used for sensitivity testing and appraisal of adaptation options. *Source:* Ranger et al. (2013)

FIGURE 14.5 Pier 4 of the Thames Barrier showing part of the mechanism required to rotate a 3300 tonne steel gate from the river bed.

of decades would be needed to secure the finance, design, construct and then operate the new infrastructure.

According to Ranger et al. (2013) the plan has four innovative features. First, rather than focusing on climate projections, the overall approach is 'decision-centric'. Understanding the vulnerability of the system, the values of the stakeholders, the decision criteria, and the options available is central to the analysis. Second, the study developed a set of plausible narratives for sea level rise based on numerical modelling and expert judgement. These enable consideration of high-end scenarios (labelled H++ in Figure 14.4) when testing the robustness of plans. Third, the plan is flexible in the face of deep uncertainty about climate change. Adaptation pathways keep multiple options open for as long as possible. Fourth, emerging climate risks are tracked using 10 indicators of changing conditions. Hence, long-term monitoring of sea level, peak river flows, intertidal habitat area, and so forth, is an integral part of the flood management system.

Case 3: Translating Science into Practice

There is no shortage of advice on how to adapt to climate change. Unfortunately, much of it is too general and/or lacks scientific underpinning. Take, for example, the view that planting trees along river banks could shade sensitive species such as salmon from lethal water temperatures (Mulholland et al., 1997). This is an appealing and common sense suggestion. However, a large number of practical details must be agreed, such as: where to plant; what species (mix); what width of planting; and what length of river bank must be covered to achieve a given reduction in water temperature? There are also many related questions about the wider ecological impact of changing local energy and nutrient fluxes to the river, flood propagation or groundwater levels. Such concerns can only be addressed through dedicated laboratory, field or modelling campaigns (Wilby et al., 2010b).

This is exactly what the Environment Agency and partners set out to achieve through the Keeping Rivers Cool project. Long-term field experiments such as the Loughborough University TEmperature Network (LUTEN) were set up to improve understanding of river section to catchment scale controls on water temperature (Johnson et al., 2014). High-resolution sampling of water temperature variations (in space and time) was undertaken alongside surveys of local conditions to determine shade effects (Figure 14.6). The information was translated into practical guidance for land managers. For

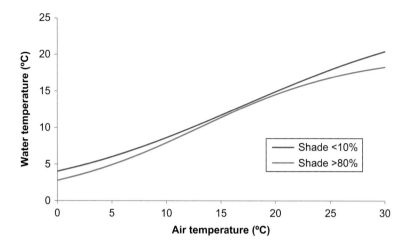

FIGURE 14.6 Modelled relationships between maximum air and water temperature at two sites in the River Dove, Derbyshire, UK. One site has less than 10% shade (red line), the other more than 80% cover (blue line).

instance, LUTEN data show that at least 1 km of tree shade is needed to reduce summer maximum water temperatures by 1 °C (Johnson and Wilby, 2015). If beech trees are planted, the assumed height of 30 m could take 80 years of growth. Faster-growing species would probably be needed to outpace climate change. Overall, shading is expected to be most effective in headwaters with little landscape shade and channel widths less than 5 m (Environment Agency, 2012).

Hallmarks of Successful Adaptation

Finally, how is it possible to tell whether any of the above adaptations will achieve intended *outcomes*? Adger et al. (2005) assert that because adaptation operates at different spatial and societal scales, success criteria should also be scale specific. Clearly, it is impossible to measure adaptation benefits before they happen unless within a modelled world (as in the Denver Water case study). The alternative is to measure aspects of the adaptation *process*. Organisations that are 'adapting well' should exhibit traits such as objective setting, routine climate risk and vulnerability assessment, and have climate awareness embedded within operations and planning (Wilby and Vaughan, 2011). On a countrywide level, the UK Committee on Climate Change (2015c) calls for the development of key indicators to track progress against the 31 objectives of the National Adaptation Plan. The next chapter imagines what could be done globally or regionally if mitigation and adaptation efforts are unsuccessful in containing climate threats.

14.2 GROUP EXERCISE: ADAPTING TO FLASH FLOODS

NOTES

This activity extends the approach to climate vulnerability and risk assessment developed in Chapters 7 and 9, respectively. These introduced the concepts of exposure, sensitivity, impact and adaptive capacity. The elements are now arranged as shown in Figure 14.7 to identify a set of low-regret adaptation measures to counter flash flood risk. Supplementary resources to support the exercise – including maps and an inventory of flood events – are available for Yemen from www.cambridge.org/wilby. Security matters aside, the region presents significant technical and logistical challenges to consultants. Data limitations combined with the complex terrain, climate regime and socio-political context must all be taken into account.

The group should be briefed on the rationale for the exercise and the severity of the flash flood hazard in Yemen (or alternative area). During each year in the period 1989–2010 flash floods claimed on average over 65 lives and displaced more than 13,000 people across affected areas of about 100,000 km^2. Agencies such as the United Nations International Fund for Agricultural Development (IFAD) are seeking to enhance the resilience of

cont.

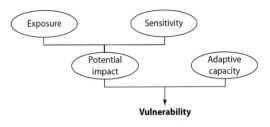

FIGURE 14.7 A conceptual framework for defining vulnerability. *Source*: Verner (2013)

communities to this serious hazard, being mindful that flood risk could increase in the future due to more intense precipitation events. Investments in local management measures can be prioritised through 'hotspot' analyses of flood risk (top-down assessment) combined with ground-truthing and community engagement (bottom-up co-production).

The group should be divided into sets of two to four students who will refer to the supporting maps and data. The exercise is divided into five tasks:

1. Evaluate the *data needs* and potential sources of information for undertaking a national flash flood risk assessment in Yemen. Groups should list the data types along with any concerns that they might have about data reliability.
2. Refer to the maps and data then determine where *exposure* to flash flood risk is presently greatest in Yemen.
3. Refer to the inventory of notable floods then determine which citizens are most *sensitive* to flash flooding in Yemen.
4. Suggest indicators of flash flood *impacts* in Yemen.
5. Identify low-regret measures that could improve *adaptive capacity* to flash flooding in Yemen (now *and* in the 2050s).

LEARNING OUTCOMES

- To appreciate that uncertainty in climate change projections is not an insurmountable obstacle to anticipatory adaptation action.
- To recognise that the feasibility and likely success of adaptations must take into account the scale-dependent socio-cultural-economic context.
- To understand that scientific evidence is needed to translate adaptation objectives into guidance for practitioners.

14.3 FURTHER QUESTIONS TO RESEARCH AND DISCUSS

What are the most significant physical, social, economic and technological obstacles to adaptation faced by coastal communities in both rich and poor countries?

Explain why particular attention is given to enabling adaptation to climate variability and change in the least developed regions of the world.

How can the success of adaptation be measured?

14.4 FURTHER READING

Carter, J.G., Cavan, G., Connelly, A., et al. 2015. Climate change and the city: building capacity for urban adaptation. *Progress in Planning*, **95**, 1–66.

Murphy, C., Tembo, M., Phiri, A., Yerokum, O. and Grummell, B. 2016. Adapting to climate change in shifting landscapes of belief. *Climatic Change*, **134**, 101–114.

Suckall, N., Stringer, L.C. and Tompkins, E.L. 2014. Presenting triple-wins? Assessing projects that deliver adaptation, mitigation and development co-benefits in rural Sub-Saharan Africa. *Ambio*, **44**, 34–41.

14.5 OTHER RESOURCES

Dartmouth Flood Observatory www.dartmouth.edu/~floods/Archives/index.html [accessed 16/07/16]

Deltares Adaptation Pathways Generator https://publicwiki.deltares.nl/display/AP/Pathways+Generator [accessed 16/07/16]

Keeping Rivers Cool Guidance www.asfb.org.uk/keeping-rivers-cool-new-guidance-for-river-managers/ [accessed 16/07/16]

Loughborough University TEmperature Network (LUTEN) www.luten.org.uk/home [accessed 16/07/16]

The Thames Barrier www.gov.uk/guidance/the-thames-barrier [accessed 16/07/16]

15

Could or Should Humankind Geoengineer Earth?

TOPIC SUMMARY

Geoengineering is defined here as the purposeful control of climate to counter the inadvertent consequences of anthropogenic GHG emissions. Hence, geoengineering may be likened to a medicine that is taken once (dangerous) symptoms appear. Technologies that enhance carbon dioxide removal from the atmosphere (e.g. biochar and land afforestation) may be relatively benign but there is uncertainty about the permanence and effectiveness of the carbon sequestered to natural sinks. Solar radiation management (e.g. via stratospheric aerosols or surface brightening) is riskier because of the potential for significant side-effects on regional climates. However, the ethical challenges posed by geoengineering deployment are generally regarded as more problematic than the technical obstacles. Even with international governance it is hard to imagine how multi-lateral consensus would ever be reached about the optimal climate conditions for all humanity. Some fear that market forces could put planet-changing technologies beyond the reach of democratic processes.

BACKGROUND READING

Schneider (1996) addresses some important technical (could) and ethical (should) questions surrounding deployment of geoengineering. Vaughan and Lenton (2011) review the effectiveness and possible side-effects of geoengineering using model,

cont.

experimental and observation evidence, whilst acknowledging the key uncertainties of each option. The Royal Society (2009) provides a systematic evaluation of geoengineering options based on their effectiveness, timeliness, safety and cost.

15.1 PLAN B
A Bitter Pill to Swallow?

What if the transition to a low-carbon global economy is too slow? What if the severity of climate change impacts overwhelms capacities to adapt? What then are the last lines of defence for society? These are uncomfortable but valid questions. To some, one answer is just as unsettling – deliberate climate control, through specified geoengineering projects. These technologies are largely untested antidotes for the climatic symptoms of rising concentrations of GHGs.

Unsurprisingly, the topic of geoengineering is highly divisive. Given that the deep cuts in emissions needed to avoid dangerous climate change may not materialise in time, it is prudent to weigh up contingency plans (see Boucher et al., 2009; Pielke et al., 2008). Some experts are totally unconvinced, believing "*geoengineering may be a bad idea*" (Robock, 2008) and that even the "*people who work on this don't want to work on it*" (Amos, 2015). Others accept deployment of such technologies as almost inevitable – "*a last gasp response to global warming*" (Morton, 2009). Still others assert that, if deliberate manipulation of the atmosphere mimics natural cooling by volcanic eruptions, the approach "*presents minimal climate risks*" (Wigley, 2006). One learned institution cautioned that climate control via geoengineering is "*not an alternative to greenhouse gas emissions reductions*" but might have some potential so warrants further research (Royal Society, 2009).

This chapter continues with a synopsis of the two main groups of geoengineering technology as defined previously: carbon dioxide removal and solar radiation management (Figure 15.1). The circumstances for and against their deployment, including considerations of legality, governance and ethics are then addressed. The concluding section reflects on the possible international dimensions and ends by questioning whether geoengineering is even compatible with liberal democracy!

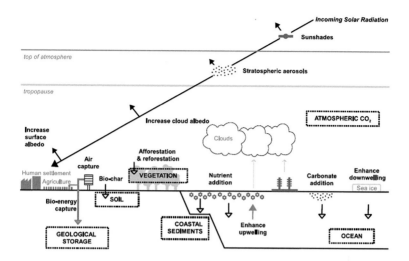

FIGURE 15.1 Schematic of main climate geoengineering proposals. *Black arrowheads* indicate short-wave radiation, *white arrowheads* indicate enhancement of natural flows of carbon, *grey downward arrow* indicates engineered flow of carbon, *grey upward arrow* indicates engineered flow of water, *dotted vertical arrows* illustrate sources of cloud condensation nuclei and *dashed boxes* indicate carbon stores. Not to scale. *Source:* Vaughan and Lenton (2011)

Carbon Dioxide Removal

This family of geoengineering technologies treats the root causes of climate change by withdrawing CO_2 from the atmosphere and transferring it to secure reservoirs within the global carbon cycle (Table 15.1). Carbon dioxide removal (CDR) may be achieved by (i) enhancing natural processes (e.g. planting forests, fertilising oceans, accelerating weathering rates) or (ii) by entirely artificial means (e.g. carbon capture and storage (CCS) at source, CO_2 capture from ambient air via chemical adsorption). Land use management to protect existing carbon sinks (e.g. avoided deforestation) also falls within the spectrum of CDR techniques. However, there are two overarching concerns about CDR: (i) the *permanence* of the carbon storage (i.e. the possibility that CO_2 might leak back into the atmosphere from biomass, soil, marine or geological reservoirs); plus (ii) the *sensitivity* of the engineered carbon balance of marine and terrestrial ecosystems to climate-related drought, fire, wind throw, pests and disease (Heimann and Reichstein, 2008). Fears about (ii) are supported by

TABLE 15.1 Descriptions of major geoengineering technologies. Adapted from Royal Society (2009)

Method	Description
Carbon dioxide removal	
Afforestation	Planting new forests on land that has not previously had tree cover in order to sequester carbon (also includes avoided deforestation schemes).
Bio-energy with CO_2 capture and sequestration (BECS)	Harvested biomass is used for fuel with capture and sequestration of CO_2 emissions, thereby drawing carbon from the atmosphere whilst generating energy.
Biomass or biochar (charcoal) carbon sequestration	Carbon fixed by biomass is stored in soils or buried where decomposition cannot return it to the atmosphere; carbon is tightly bound as charcoal in soils.
Carbon capture and storage (CCS) at source	Collecting CO_2 emissions from large point sources then sequestering in geological strata such as offshore/onshore saline aquifers, depleted oil and gas reservoirs, un-economic coal seams, shales and cavities.
CO_2 air capture	Processes that draw CO_2 from ambient air via adsorption onto ion-exchange resins, into highly alkaline solutions, or into moderately alkaline solutions with a catalyst.
Enhanced weathering	Natural mechanisms by which carbon is drawn from the atmosphere via the dissolution of carbonate and silicate rocks enhanced by applying olivine to agricultural soils or by 'liming' the ocean.
Ocean fertilisation	Addition of otherwise limiting nutrients (such as nitrogen, phosphate or iron) to surface

cont.

TABLE 15.1 (cont.)

Method	Description
	waters to promote algal growth, faster removal of CO_2 from air and the fraction sequestered to deep water or ocean sediments.
Ocean upwelling or downwelling	Enhanced transfer of atmospheric carbon to deep ocean by increasing supply of nutrients using vertical pipes to pump water from depths to the surface or by promoting downwelling of dense water in subpolar oceans.
Solar radiation management	
Cloud albedo	Whitening clouds over the ocean to reflect more solar radiation by increasing the number of cloud condensation nuclei (sea salt particles) in low-level marine clouds, using a fleet of vessels.
Space reflectors	Placement of dust particles, metallic mesh, swarms of reflecting or refracting disks in near-Earth orbits to reflect or deflect solar radiation back to space.
Stratospheric aerosols	Release of aerosols into the stratosphere to scatter sunlight back to space to mimic the effect of large volcanic eruptions on the global climate.
Surface albedo (rural)	Selection of crop varieties and grasslands that are more reflective; covering deserts with reflective polyethylene–aluminium surfaces; reforesting tropics and sub-tropics to cool surfaces via increased evaporation and transpiration.
Surface albedo (urban)	Local brightening of urban areas via white roofs, roads and pavements to reflect more solar radiation.

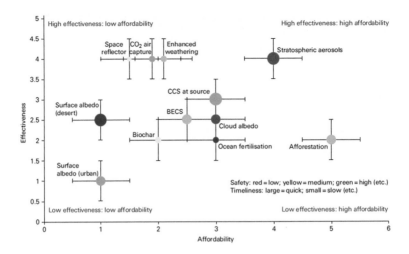

FIGURE 15.2 Geoengineering effectiveness versus affordability, with the size of points indicating timeliness (larger if more rapidly implementable and effective), and colour indicating safety (green if safe, through to red if not) of different methods. *Source*: Royal Society (2009)

empirical evidence showing that the fraction of CO_2 emissions remaining in the atmosphere may have increased since the 1950s due to climate weakening of carbon uptake by ecosystems (Le Quéré et al., 2009).

Effectiveness, affordability, timeliness and safety need to be judged for each CDR technique (Figure 15.2). Land management, biomass harvesting and afforestation may have relatively low risks of unanticipated effects; but the amount/pace of carbon sequestration may be too low, and land set aside for geoengineering may be land lost to food production (Vaughan and Lenton, 2011). Biochar ploughed into agricultural soils enhances nutrient retention; but the charcoal could darken earths and thereby increase local warming and soil respiration rates. Carbon dioxide air capture and accelerated weathering may be relatively safe and flexible options for counter-acting small/mobile emission sources, but their costs and technical feasibility are uncertain (Keith, 2009; Stephens and Keith, 2008). Ocean fertilisation has low cost-effectiveness and is considered to have high potential for undesirable ecological side-effects, such as enhanced respiration levels by marine algae depleting oxygen levels and expanding anoxic regions ('dead zones') in the ocean (Royal Society, 2009). Ocean acidity would be expected to increase too.

According to the Global Carbon Capture and Storage Institute, there were 15 large projects in operation by 2015 and another 7 under construction with combined capture capacity of ~40 million tonnes of CO_2 per annum (Mtpa), equivalent to the total emissions of Denmark in 2014. By 2016, a further 18 major projects had reached the planning stage with the potential to capture another 25 Mtpa (equal to all the emissions of Tunisia). The technology has appeal because it enables continued use of fossil fuels while reducing emissions. However, CCS experienced an unexpected setback in 2015 when the UK Government cancelled pilot projects (House of Commons, 2015). Such political reservations may reflect concerns about the uncertain pathway to commercial deployment and up-scaling of the technology (see Haszeldine, 2009). Others claim that financial explanations are too simplistic and that deployment (or not) of CCS demonstration projects has to be seen in the context of the wider political economy (Kern et al., 2016). For instance, strong uptake in Canada may be tied to projects to unlock the vast oil sand reserves of Alberta.

Solar Radiation Management

Solar radiation management (SRM) technologies are more controversial than CDR because they do not address the root cause of global warming and could actually exacerbate some symptoms such as ocean acidification or regional droughts. Moreover, if for some reason the deployment was halted, rapid warming would return. This is because all such schemes attempt to rectify the enhanced radiative forcing due to rising concentrations of anthropogenic GHGs *without* directly reducing the amount of those gases in the atmosphere (Table 15.1). Surface brightening of urban landscapes is relatively low risk and has the added benefit of reducing urban heat islands, but would be least appealing in terms of total global effectiveness and affordability (Figure 15.2). Methods such as reduced tillage could enhance the albedo of agricultural soils and attenuate peak temperatures during heatwaves whilst retaining more moisture and carbon in soils (Davin et al., 2014). However, the cooling effect of urban and agricultural surface albedo modification is strongly seasonal, most effective under clear-sky conditions, and largely restricted to the areas of deployment. Increasing the reflectivity of large expanses of desert would be more effective, but takes the land out of other productive uses and carries a host of adverse environmental side-effects, not least potential disruption of the Indian and African monsoons (Irvine et al., 2011).

Early proponents of methods that increase the albedo (i.e. reflectivity) of the atmosphere point to the sulphate loading over industrial areas and abrupt cooling caused by volcanic eruptions such as Pinatubo in 1991 (Crutzen, 2006; Wigley, 2006). This single event injected an estimated 10 million tonnes of sub-micrometre SO_2 particles into the stratosphere, which back-scattered solar radiation to space and cooled the Earth's surface by 0.5 °C during the following year. Other evidence of the expected residence time, global transport and ultimate fate of stratospheric particulates is provided by data from nuclear weapons testing in the 1950s and 1960s (Tuck et al., 2008). Opponents of stratospheric geoengineering list a range of concerns (Table 15.2), many of which are due to potentially undesirable or unanticipated effects on human and natural systems. For example, Xia et al. (2014) predict significant reductions in rice yields across northern China under simulated SRM scenarios.

With the advent of bespoke modelling experiments it is possible to investigate other direct and indirect effects of SRM (Tilmes et al., 2015). For example, one assessment modelled the temperature and precipitation responses of six different SRM schemes (Crook et al., 2015). They found that amongst the options that potentially restore future surface air temperatures to those of 1986–2005 (ocean and stratocumulus albedo enhancement, cirrus cloud thinning and stratospheric SO_2 injection), all resulted in large-scale changes in tropical precipitation. Furthermore, real-world implementation would be problematic because of the non-linearity and asymmetry of responses between hemispheres (especially under desert albedo enhancement). Other studies reveal that SRM techniques would be ineffectual at preventing mass loss from the Greenland and West Antarctic ice sheets and hence would not prevent sea level rise (Applegate and Keller, 2015; McCusker et al., 2015). Likewise, Gabriel and Robock (2015) were unable to detect any change in ENSO event frequency or amplitude in a geoengineered world when compared with historical or unfettered global warming simulations.

The Case for Geoengineering Research

There is clearly a range of arguments for and against all geoengineering methods. However, the Royal Society (2009: xi) assert: *"The greatest challenges to the successful deployment of geoengineering may be*

TABLE 15.2 Twenty reasons for concern about climate control via injection of sulphate aerosols into the stratosphere. Adapted from Robock (2008)

1. Unanticipated or undesirable effects on regional wind and precipitation patterns

2. Solar radiation management does not address continued ocean acidification

3. Injected aerosols could destroy stratospheric ozone

4. Plants affected by reduced direct solar radiation and increased diffuse radiation

5. Increased acid deposition causes harm to ecosystems and human health

6. Poorly understood effect of particles on cirrus cloud formation in the troposphere

7. Perpetual whitening of the sky (but colourful sunsets)

8. Reduced energy production from solar power systems

9. Environmental impacts of aerosol delivery systems such as fired shells or jets

10. Rapid warming if deployment is halted by social, political or technological crisis

11. Uncertainty about rate of reversibility in the event of too much cooling

12. Human error(s) in design and delivery of the system (by lowest bidder)

13. Perceived as a 'get-out-of-jail free' card that hinders mitigation efforts

14. Uncertainty about costs relative to safer alternatives offered by renewables

15. Control of technology ceded to commercial entities and their shareholders

16. Potential for weaponisation of geoengineering technologies generally

cont.

TABLE 15.2 (cont.)
17. Conflicts with existing international treaties and legislation
18. Difficulty in agreeing the optimal climate
19. Questions around the moral legitimacy of deliberate climate control
20. Unexpected consequences arising from complex socio-climate interactions

the social, ethical, legal and political issues associated with governance, rather than scientific and technical issues." What then might be the special circumstances under which these non-technical concerns could be over-ruled? MacCracken (2009) identifies three red flags that could wave on the implementation of geoengineering. All invoke climate impacts on society that are so severe that even the cost and inter-generational risks of SRM might be countenanced. His scenarios are (i) warming of low-latitude oceans leading to more intense high-category tropical cyclones, droughts and coral bleaching; (ii) warming of high latitudes with associated loss of Arctic sea ice, destabilisation of the Greenland and Antarctic ice sheets; and (iii) loss of the regional cooling effect of SO_2 emissions linked to reductions in CO_2 emissions from coal-fired power plants.

Others have assessed the feasibility of geoengineering specific extreme weather hazards. For example, Moore et al. (2015) investigated the possibility of reducing the frequency of Katrina-like hurricanes in the Atlantic using stratospheric aerosols to preferentially cool the tropics relative to polar regions. They modelled an SO_2 dosage equivalent to Pinatubo once every 2 years to balance temperature increases under RCP4.5 and found a 50% reduction in the number of hurricanes compared with no geoengineering. Expected 5-year storm surge levels were 40 cm lower because of the reduced amount of sea level rise. Similarly, Ghosh et al. (2016) determined that seeding sea-salt droplets in the eye of a tropical cyclone over the Bay of Bengal could decrease heavy rainfall by up to 50% upon landfall. Other suggestions for regional or 'soft' geoengineering include the use of micro-bubbles to increase reflectivity and cool the ocean near threatened coral reefs (Olson, 2012); or use of hydrosols to whiten and cool Arctic wetlands, thereby stemming the release of the potent GHG methane (Seitz, 2011). However, there is still large uncertainty about possible global side-effects even from regional geoengineering.

An international framework of governance for geoengineering research has yet to emerge. Some groups have tried to fill this void by developing consensus statements. For example, the Oxford Principles of Geoengineering (Rayner et al., 2009) set out five doctrines for overseeing such research: (i) establish governance architecture before deployment; (ii) provide independent assessment of impacts; (iii) disclose research results; (iv) engage the public in the decision-making process; and (v) regulate as a public good. Some research would be covered by existing international law, such as the UN Law of the Sea Convention, which prohibits activities in one territory that could harm other states. But it is unclear whether scientific self-regulation and laws would be effective in the case of privately funded or rogue projects. Parson and Keith (2013) propose that shortcomings could be addressed by (i) accepting government authority and international coordination of geoengineering research; (ii) declaring a moratorium on large-scale geoengineering (above a threshold radiative forcing); and (iii) applying modest levels of regulatory scrutiny and transparency to small-scale interventions. Their overall intent is to enable legitimate research to proceed whilst building norms of shared knowledge, cooperation and transparency at a formative stage of the work.

Deployment of Geoengineering

Even if a scientific case can be made and technologies are developed, there are still profound ethical questions about geoengineering deployment. Some fear that geoengineering could divert attention and resources away from mitigation and adaptation efforts, or lessen the incentive to act against the causes of climate change if society believes that it is protected from harm by geoengineering (i.e. moral hazard) (Table 15.2). Cicerone (2006) suggests that scientists should support research into geoengineering but stop short of participating in deployments. MacCracken (2009:10) asks: "*What if geoengineering is started and it does not work as expected — what is irreversible and what is not? Does geoengineering lead to winners and losers? What are the optimal conditions for the Earth — and, if they exist, would they simultaneously be optimal for all peoples, for society, and for plants and wildlife? Who would get to decide? Once geoengineering started, how long would it have to continue? How soon do decisions about geoengineering have to be made? Who would pay for and carry out the geoengineering efforts?*"

The international ramifications of geoengineering cannot be ignored. As mentioned above, the benefits (and possible harms) would be unevenly distributed in space and time. There is a real risk of unilateral action using cheap technologies, such as the controversial marine fertilisation project that took place west of British Columbia in 2012 (Parson and Keith, 2013). Such actions could threaten international solidarity or compromise other areas of climate policy. Above all, there is uncertainty about jurisdiction and whose hand would be on the [global] thermostat (Robock, 2008).

Although public acceptability cannot be taken for granted (Corner and Pidgeon, 2010), some geoengineering technologies (e.g. SRM) could actually fall outside democratic processes and institutions (Szerszynski et al., 2013). This is because of the impossibility of informed consent when the aims and outcomes of deployment are so heterogeneous across different sovereign states. Involvement of transnational corporations and the private sector could push technologies beyond the regulatory powers of individual governments. Unfortunately, patents for geoengineering are already beginning to place control of climate-changing technologies in the hands of a few people (Cressey, 2012). Most worrying, patent holders would have a vested financial interest in keeping the climate in a state of near-catastrophe. Perhaps the most urgent question is not whether we could or should geoengineer Earth, but whether citizens will ever fear climate change enough to relinquish national self-determination to remote, autocratic structures?

15.2 GROUP EXERCISE: GLOBAL GEOENGINEERING COUNCIL

NOTES

This activity explores important technical and governance issues surrounding the use of geoengineering. The session begins with an imagined scenario in which some of MacCracken's (2009) conditions for deployment are met – but other scenarios could be used. The scene-setting is followed by a semi-structured group discussion on matters of target climate, rules of engagement and geoengineering governance. Table 15.1 and Figure 15.2 could be provided for reference. There are no set answers but some suggestions are offered to guide the debate.

A Scenario: The year is 2030 and the global community has failed to build on the emission reductions pledged at the

cont.

2015 Paris COP. In the past decade, catastrophic storm surges have struck several coastal megacities. Deadly heatwaves are now occurring almost annually in the United States and South Asia. Parts of the West Antarctic Ice Sheet have begun to disintegrate. As a member of the independent Global Geoengineering Council (GGC) you are now seriously debating the following issues:

1. What target climate is the GGC proposing to achieve?
2. Linked to (1), what geoengineering technologies does the GGC recommend?
3. What should be the 'rules of geoengineering deployment'?
4. What body should govern the geoengineering deployment?
5. Who should be invited to join the GGC Advisory Group?

LEARNING OUTCOMES

- To appreciate that various geoengineering methods can be compared in terms of their cost-effectiveness (relative to conventional mitigation techniques), safety, timeliness, reversibility, possible unintended impacts and public acceptability.
- To be aware of the major ethical considerations surrounding geoengineering deployment such as the uneven pattern of benefits (and harms), optimal level of climate control, and scope for moral hazard.
- To recognise that whilst no international governance structures are yet in place to regulate geoengineering research and deployment, some technologies have already been patented.

15.3 FURTHER QUESTIONS TO RESEARCH AND DISCUSS

How might the benefits of geoengineering deployment(s) be assessed?

Evaluate the relative merits of an international framework versus grass-roots regulation of geoengineering research.

What are the dangers of growing private sector involvement in geoengineering research and deployment?

15.4 FURTHER READING

Caseldine, C. 2015. So what sort of climate do we want? Thoughts on how to decide what is 'natural' climate. *Geographical Journal*, **181**, 366–374.

Keith, D.W. and MacMartin, D.G. 2015. A temporary, moderate and responsive scenario for solar geoengineering. *Nature Climate Change*, **5**, 201–206.

Stilgoe, J. 2015. *Experiment Earth: Responsible Innovation in Geoengineering*. Routledge, London.

15.5 OTHER RESOURCES

Climate Action Tracker http://climateactiontracker.org/ [accessed 16/07/16]

Climate Geoengineering Governance (CGG) project http://geoengineering-governance-research.org/ [accessed 16/07/16]

Global Carbon Capture and Storage Institute www.globalccsinstitute.com/ [accessed 16/07/16]

Integrated Assessment of Geoengineering Proposals (IAGP) http://iagp.ac.uk/ [accessed 16/07/16]

National Academy of Sciences: Climate Intervention Reports http://nas-sites.org/americasclimatechoices/public-release-event-climate-intervention-reports/ [accessed 16/07/16]

Panel discussion of geoengineering. Woodrow Wilson International Center for Scholars www.youtube.com/watch?v=lvsu0AC0t-w [accessed 16/07/16]

16

How Is Climate Change Communicated?

TOPIC SUMMARY

To raise awareness and shape attitudes to climate change issues, there needs to be a sound process of communication between the scientist and the public. This can be done by (i) writing scientific reports in journals and newspapers and hoping that the public will pick up key messages, or (ii) actively engaging the public and scientist in debate where outcomes are contextual, co-produced and continually reshaped. Mass media intervene in both cases, to frame the debate according to prevailing ideologies and intentions. Common frames for climate change include catastrophism, uncertainty, security, injustice, economic or biodiversity impacts. Each engenders different reactions in the audience, such as a sense of hopelessness, resistance to change, fear, motivation to act, self-interest or concern. Socio-cultural factors and personal circumstances determine the ways in which these messages are received and the extent to which audiences are empowered to act either individually or collectively.

BACKGROUND READING

Hulme (2009) explains why there are multiple and often conflicting messages about climate change coming from different media and sources. Moser (2010) reviews the many obstacles faced by communicators when trying to convey the issue of anthropogenic climate change, as well as those aspects of the process that determine effectiveness of communication.

16.1 MIXED MESSAGES
Climate Spin

Ed Hawkins' animated spiral of global mean surface temperature change has been making quite a 'stir' (Figure 16.1). Rather than plotting global warming as a conventional line chart (Figure 1.1), his graphic shows the same data in a more eye-catching way. The spiral begins in 1850, expanding outwards as the climate warms or spinning inwards during cooler interludes. Each cycle marks a calendar year with colours representing the passage of time from purple, through blue, green and yellow for recent years. The viridis colour scale was chosen because it works for the colour blind and has even contrast of shading across its range (Hawkins, 2015a). Periods of the animation that stand out are the 1880s–1910 (modest cooling); 1910–1940s (warming); 1950s–1970s (stable temperatures); 1980s onwards (strong warming). A marked jump in April 2015 (partly due to a strong El Niño) throws the monthly mean global temperature anomaly close to 1.5 °C. Seen in this format, even the 2 °C marker for 'dangerous' climate change looks perilously close.

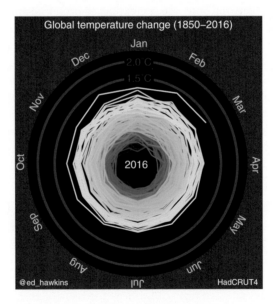

FIGURE 16.1 Spiralling global mean temperatures. For the animated version see www.climate-lab-book.ac.uk/files/2016/05/spiral_optimized.gif
Source: Ed Hawkins

Climate change can be communicated in ways that either mobilise or stall action. Some groups look at the evidence and are persuaded that human civilisation has reached a critical moment and call for more effective sharing of information about potential climate risks and opportunities (Bowman et al., 2010). Others seek to undermine public trust in science, thereby discrediting the messengers and delaying decision-making – they choose to ignore or interpret the same data in different contexts (see McCright, 2010). The relative credibility of these competing voices is examined in Chapter 17, but it is important to appreciate that the messenger and the message are inseparable elements of the climate communication process. Above all, how the message is presented or framed really matters (Kuntzmann and Drake, 2016). According to Entman (1993:52): *"To frame is to select some aspects of a perceived reality and make them more salient in a communicating text, in such a way as to promote a particular problem definition, causal interpretation, moral evaluation, and/or treatment recommendation."*

The rest of this chapter presents a general account of the evolving approach to science communication: from where diverse public audiences are treated as passive receivers of information, to where they are now partners in scientific discourse. Communicators still face many challenges and some of the most important obstacles will be covered too. Key aspects of the communication process are then described, including the repertoires, framing and channels of message delivery. The closing section considers the effectiveness of climate communication in changing behaviour.

Evolving Models of Science Communication

Research into science communication can be organised around actors such as the public, scientists, journalists and news organisations (Weigold, 2001). Opinions vary about the level of scientific literacy of the public, but the assumption of knowledge deficiency is often implicit in the communication process. Ignorance is problematic for the public and scientist alike. The former are vulnerable to anti- or pseudo-science messages; the latter depend on public goodwill and support for their funding. The 'deficiency' model is top-down and science-led – it assumes that clear messaging, handled by neutral media, heard by receptive audiences, raises the 'public understanding of science' (Figure 16.2a). In this model, there may be a temptation to 'shout louder' about the risks to counter apparent

(a)

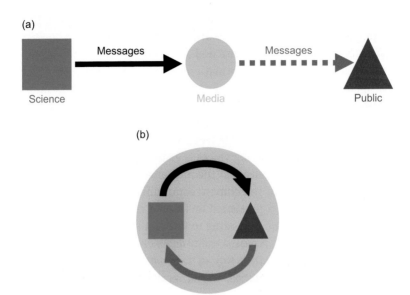

(b)

FIGURE 16.2 Conceptualisation of the relationships between science, media and the public in (a) deficit and (b) cultural circuit models of communication.

inaction. Arguably, this has been the case with the headline messages of the IPCC. As the tone of successive assessments and media reports has become more sombre, some fear that this has led to climate fatigue (Kerr, 2009), denial or even counteractive behaviour (Hulme, 2007).

There are other problems with the deficiency model. First, the (global) audience is not uniform in terms of either scientific literacy or attentiveness to scientific issues. One international study found that 80% of 15-year-old students were defined as scientifically literate, with the highest average scores in Finland but below average (for OECD countries) in France, Spain and the United States (Bybee et al., 2009). Second, as will be explained below, the media frame, filter and interpret messages to fit their own ideology and/or perceived interests of their audiences – the mass media are not neutral purveyors of knowledge. Nor are they particularly well informed themselves. For instance, the House of Commons Science and Technology Committee (2014) found that the leading public service broadcaster in the UK (the BBC) "*lacked a clear understanding of the information needs of its audience with regards to climate science*". Third, more nuanced versions of the deficiency model ask what do the public *need* to know to be good

FIGURE 16.3 The M1 motorway and River Trent, UK, during the November 2000 floods.

citizens or what would people *want* to know in their circumstances (Weigold, 2001). Both require value judgements to be made by the information supplier. Finally, there is evidence from the United States that those who are more scientifically informed feel less concern and responsibility for global warming (Kellstedt et al., 2008).

Advocates of the 'cultural circuits' model claim that the producers and consumers of media products are joined in dialogues about climate change that are dynamic and context-specific. In this model, the media are not conduits of knowledge between remote senders (scientists) and receivers (the public); rather the media create arenas within which messages are continually shaped (Figure 16.2b). For example, it is claimed that growing awareness of dangerous climate change (prompted in part by severe flooding in 2000 in the Midlands (Figure 16.3) and the deadly European heatwave in 2003) helped to soften UK media and public attitudes to nuclear energy as an effective mitigation option (Carvalho and Burgess, 2005). Discourse is also stimulated by collaborative projects such as *climateprediction.net*, which enables mass participation of the public in climate simulation experiments (Chapter 3). Other 'citizen science' initiatives like *Old Weather* involve the public in transcribing Arctic weather observations held in ships' logs into digital format.

Not everyone is entirely convinced by the merits of this circular model of communication and co-production of knowledge. In an era of 'big data', some feel that the value of information not originally intended for research purposes has been grossly overrated because of biases, measurement errors and misclassifications, to name a few (Ioannidis, 2013). From a post-modern perspective there is also the possibility of multiple, conflicting interpretations of the same data when scientific truth is regarded as relative rather than absolute (Verosub, 2010). Climate change messages from expert and lobby groups can become blurred, as was the case after the exceptionally warm year 2015. NASA/NOAA greeted the occasion with the headline[1] *"Analyses reveal record-shattering global warm temperatures in 2015"*; whereas the contrarian Global Warming Policy Foundation[2] chose *"2015 global temp, or how some scientists deliberately mistook weather for climate"*. In practice, the extended conversations needed to align such divergent ideas can only feasibly occur between a few scientists and unevenly attentive/engaged lay audiences.

Climate Communication Processes

According to Moser (2010) there are three reasons for climate change communication: (i) to inform and educate; (ii) to achieve social engagement and action; and (iii) to revise social norms and cultural values. The approach may be calibrated according to the target audience using different framing, language, conveyor(s), and mode(s) of message communication. For example, different sets of images and repertoires would be suitable for primary school pupils compared with university students, yet surprisingly little research is available on communicating climate change to children. Fear-inducing (disaster) messages may draw attention to the issue but seldom translate into changed behaviour (Lowe et al., 2006; O'Neill and Nicholson-Cole, 2009). Such frames would be inappropriate in a junior classroom; more constructive methods of engagement include online carbon calculators, teaching activities and games (see online resources). These all gently build scientific understanding whilst promoting personal involvement.

[1] www.nasa.gov/press-release/nasa-noaa-analyses-reveal-record-shattering-global-warm-temperatures-in-2015
[2] www.thegwpf.com/2015-global-temp-or-how-some-scientists-deliberately-mistook-weather-for-climate/

Overall, the frame is the most important aspect of the communication process because it shapes the choice of repertoire, imagery, symbolism, non-visual cues and conduits of delivery. More generally, the vernacular of climate has been evolving. What was once narrowly described as 'global warming' has morphed into more inclusive ideas as 'climate change', 'global climate disruption', 'climate chaos' or 'climate challenges'. Depending on the frame, such words can convey consecutively a narrow definition (as benign warmth only), a broader but unspecified suite of climate threats, a growing sense of imminent catastrophe, or an opportunity to conquer the future climate (Hulme, 2008; Risbey, 2008). According to the pragmatic climate repertoires identified by Ereaut and Segnit (2006), UK party political language ahead of the 2010 General Election (Table 16.1) could be characterised as 'small actions/corporate' (Conservatives), 'David versus Goliath' (Liberal Democrats), or 'techno-optimism' (Labour). By the 2015 election, major differences in the framing of climate change by the major parties were still self-evident. For example, the domestic agenda of the Scottish National Party contrasted with the global endeavour voiced by the Liberal Democrats (Table 16.2).

TABLE 16.1 UK political party repertoires on climate change prior to the 2010 election

Conservatives	Liberal Democrat	Labour
Challenge	Scientists	Importance
Moral	Last chance	Issue
Act	Bold measures	Investment
Cooperation	Path	Make a difference
Tough regulation	Green	Build
Performance	Sustainable	Low carbon
Vision	Growth	Renewables
Big society	Differently	Cooperation
Choice	Challenge	Lead
Driving force	Trust	World deal
Sustainable	Record	Together
Green investment	Conviction	Changes
Encourage	Communities	Green
Optimism	International	Environment
Urgency	Together	Revolution
Opportunity	Opportunity	
Massive impact		

TABLE 16.2 UK political party framing of climate change prior to the 2015 election. *Source*: Carbon Brief www.carbonbrief.org/election-2015-what-the-manifestos-say-on-climate-and-energy [accessed 25/06/16]

Conservatives	*"We will cut emissions as cost-effectively as possible, and will not support additional distorting and expensive power sector targets."*
Green Party	*"Climate change is the greatest challenge of our time and only the Greens are determined to tackle it."*
Labour	*"…tackling climate change is an economic necessity and the most important thing we must do for our children, our grandchildren and future generations."*
Liberal Democrats	*"Climate change, one of the greatest challenges of our age, is by its nature global."*
Scottish National Party	*"We will use our influence at Westminster to ensure the UK matches, and supports, Scotland's ambitious commitments to carbon reduction."*
UK Independence Party	*"[The UK's] failing energy policies… will do nothing to reduce global emissions."*

Frames also betray the ideological positions held by the media (e.g. Doulton and Brown, 2009). These include scientific uncertainty (as a right-wing brake on decision-making and action), business as usual (liberal market options), through to economic reform to decarbonise economies (via highly interventionist, left-wing policies). An early study of the frames used by UK tabloid newspapers found stories about climate change were often told in relation to political actors, endangered animals or extreme weather events (Boykoff, 2008a). Quantitative analysis of news content in the United States revealed that 70% of segments provided balanced coverage of 'natural' versus 'human' causes of climate change, thereby promoting messages of scientific contention rather than convergence (Boykoff, 2007, 2008b). Elsewhere, alarmism and nihilistic humour are used by journalists to respectively exaggerate or trivialise threats posed by climate change (Table 16.3). More specifically, framing using natural analogies (carbon uptake by trees and cooling by volcanic particles) can be employed to increase public positivity towards climate geoengineering (Corner and Pidgeon, 2015). [See the group exercise below for further examples of media frames.]

TABLE 16.3 Selected examples of alarmist messages and nihilistic humour about climate change in British newspapers over the period 2000–2016

Headline	Date
"It's the end of the world – mainly for children", Express	17/01/2000
"Just 100 years to go folks!", News of the World	03/02/2002
"Wave it goodbye: Raging floods could swamp our cities within a lifetime", Mirror	16/09/2002
"How the world will end", Daily Mail	28/12/2002
"Gas what! Cows get wind cure", Sun	30/04/2003
"Lights on but nobody home", Express	09/01/2006
"Snow joke", Sun	13/01/2006
"Pollution is turning the seas into acid", Daily Mail	11/11/2006
"A 21st century catastrophe", Independent	23/07/2007
"No ice at the North Pole", Independent	26/06/2008
"First came the floods, then came the snow...and the climate scientists were silent", Times	06/03/2011
"One day, turning off the lights won't be up to you", Telegraph	23/02/2013
"The next big freeze could last 250 years...", Mail	29/04/2013
"North Pole 'moving TOWARDS the UK' as climate change shifts way Earth wobbles", Mirror	09/04/2016
"Does the Committee on Climate Change want to blow us all up?", Telegraph	23/04/2016

Climate Communication Challenges

Moser (2010) provides a critique of the special challenges faced by communicators of climate change. The sheer complexity and uncertainty of the issue are surely major impediments. Another barrier is the invisibility of the underlying cause – no one can see, taste or smell carbon dioxide, unlike other atmospheric pollutants (see Figure 10.4a). Outside the laboratory, the effects of CO_2 on radiant energy and water vapour feedbacks are just as imperceptible. Moreover, the links between cause and effect may be separated by significant distances in space and time, so are only detectable through global monitoring over many decades. The earliest warning signs of changing climate are also disconnected from largely urbanised populations by virtue of the physical remoteness of fragile environments (e.g. Arctic landscapes, tropical rainforests, coral reefs, mountain ranges). Some even contend that the skills of abstraction needed to believe in the power of humans over the global climate fall outside of the evolutionary survival trait of responding to more immediate danger signals (De Martino et al., 2006).

Modern humans and their livelihoods are also insulated (to varying degrees) from the threats posed by climate variability and change. For instance, flood defences afford some protection from storms; air-conditioned buildings and vehicles mask the silent threat posed by heatwaves. Not surprisingly, those communities whose livelihoods are most vulnerable to climate shocks are also more sensitive to climate change signals. For example, rural respondents in India with high food and livelihood dependence on weather are more likely to perceive actual changes in rainfall than their urban counterparts (Howe et al., 2014). Similarly, the effect of the 2012 Midwestern United States drought increased the perception of risks from drought and pests by agricultural advisors (even though there was no significant change in their climate change beliefs or adaptation attitudes) (Carlton et al., 2016).

Direct personal experience of an extreme event can make certain climate threats more tangible to audiences (Whitmarsh, 2008). Front-line communities who are already experiencing adverse climate impacts are also more likely to be the focus of media reports and linked scientific messages about the need for urgent mitigation actions (Orlove et al., 2014). Although small island states such as Tuvalu in the South Pacific have been prominent in calls for a 1.5 °C global mean warming target, narratives of an imminent climate change crisis and exodus of refugees are contested (Farbotko and Lazrus, 2012). Under less extreme circumstances, the memory of past notable weather events can be used as analogues for the 'new normal' conditions under climate change. For example, Matthews et al. (2016) believe that by referring to the warmest and driest summer on record in Ireland (1995), the public will relate better to future drought risks.

Other obstacles to communication include the weakness of emergent signals relative to the general noise of climate variability, the immense complexity of the associated human–environmental interactions, and uncertainty of future impacts (Moser, 2010). Add to these the powerful resistive forces of short-term self-interest along with the tendency to preserve the *status quo* and one result is cognitive dissonance. This is when our behaviours contradict our attitudes or beliefs about climate change. Hence, green consumers make various excuses to justify their continued air travel despite knowledge of the harmful impact on climate (McDonald et al., 2015). In such cases, it is too simplistic to assume that better climate communication and education automatically translate into behaviour change. A study of climate-friendly actions and policies in Switzerland found that high-cost behaviours are more likely to be dismissed by the public as less effective options for climate mitigation (Tobler et al., 2012).

Effectiveness of the Communication

Given the many obstacles and complex processes involved, does climate communication actually work? What is the evidence that climate communication can educate, engage and ultimately change behaviours? Are audiences empowered to make changes in behaviour? Rising global emissions might suggest that past communication efforts have been unsuccessful. Indeed, Nerlich et al. (2010:101) state flatly that there is *"no direct correlation between communication and behaviour change"*. This is because knowledge transmission is non-linear, takes place in an open system, and is nuanced by culturally mediated variations in public perceptions of the issue. Moreover, the scope for action is determined materially by an individual's socio-economic context as well as their knowledge and attitudes. Hence, findings from laboratory-based psychological studies do not translate into neat formulas for effective climate communication. Some are calling for more concerted action to strengthen communication science itself (Pidgeon and Fischhoff, 2011).

There is a widespread assumption that pro-climate behaviour emerges from recognition of *personal* accountability for climate change. Previously, it had been thought that collective framing diminishes the incentive for individuals to act. However, recent research shows that donations and intent to reduce carbon emissions increase when *collective* responsibility is emphasised (Obradovich and Guenther, 2016). This suggests that communication campaigns should present climate change as a collective challenge to which individuals can contribute meaningful solutions. Pearce et al. (2015) also call for more open debate about uncertainty and complexity alongside consensus messages. For instance, the so-called 'global warming pause' (1998–2014) created space for discourse around the limitations of observational data (at high latitudes and for the deep ocean) and elusiveness of climate model sensitivity. At the very least, genuine acknowledgement and attention to areas of scientific ambiguity could foster more meaningful public engagement. As has been shown before, acceptance of the indeterminate nature of the future climate is an important step towards developing robust mitigation and adaptation strategies.

It is evident that the media both influence levels of public concern and provide a barometer of changing attention to climate issues. For example, Sampei and Aoyagi-Usui (2009) reported a dramatic rise in Japanese concern about global warming during 2007 that matched an increase in the total number of newspaper articles on the topic. Less is known about social media as a conduit for shaping public opinions about climate change beyond shifting the emphasis from scientific to political debate (Pearce et al., 2015). One study of Twitter messages concluded that discussions of climate change typically occur within polarised spheres of like-minded users ('echo chambers') (Williams et al., 2015). But some climate stories on social media can be picked up by news journalists and passed on to broader audiences (Schäfer, 2012).

Newspaper outlets exhibit peaks in the number of articles that coincide with global events such as the UN climate summits in Copenhagen (December 2009) and Paris (December 2015) (Figure 16.4). Greater media attention during the 2009 summit was probably due to

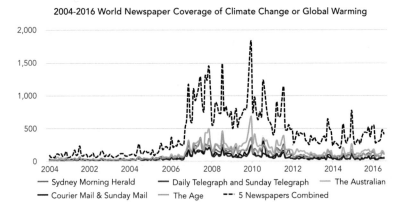

FIGURE 16.4 Number of newspaper articles covering climate change or global warming in Australia 2004–2016. *Source*: Boykoff, M., K. Andrews, M. Daly, L. Gifford, G. Luedecke, L. McAllister, and A. Nacu-Schmidt, 2016. Australian Newspaper Coverage of Climate Change or Global Warming, 2000–2016. Center for Science and Technology Policy Research, Cooperative Institute for Research in Environmental Sciences, University of Colorado, Web [accessed 19/07/16]

the direct involvement of President Obama in last minute negotiations, as well as the hacking of emails between climate scientists during the affair known as 'Climategate' (Chapter 17). There has also been a general drift away from print news to online content since 2009. With specific regard to Australia, peaks in the number of articles chart domestic concerns about the extensive flooding in Queensland and New South Wales (January 2011 onwards); the release of a new emissions trading scheme (July 2011); the incoming Abbott Government first abolishing the Climate Commission (September 2013) then repealing carbon tax (July 2014). The next chapter takes a deeper look at the expertise and role of different messengers in shaping wider policy agendas on climate change.

16.2 GROUP EXERCISE: FRAMING CLIMATE CHANGE

NOTES

Shanahan (2007) describes six frames of climate change and explains how each might appeal to different audiences. These are (i) *scientific uncertainty* (for those who do not want to change); (ii) *national security* (for those in group (i) who are now inspired to act); (iii) *loss of biodiversity* (for those who love nature); (iv) *economic impacts* (for those with political influence and/or business interests); (v) *catastrophes* or *extreme events* (for those who worry about the future); and (vi) *injustice* or *inequity* (for those with strong ethical beliefs).

This activity can be run in two parts. First, prepare about a dozen images that sample the six frames above. Other frames might be represented too, such as *humour* (for those who do not want to worry about the future) or *risk management* (for those who accept that the future is uncertain but action can reduce the likelihood of catastrophe) (see Weber and Stern, 2011). Images can be sourced from various media channels (e.g. art, cartoons and video). See, for example, Figure 16.5 or the link to Climate Change Art (see Section 16.5) for some thought-provoking material.

Images should be displayed long enough for the group to assign each to one or more frames. Ideally, some context can be given to every image after it has been discussed by the group. At the end of the slide show, the group should reflect on the extent to which Shanahan's (2007) six frames are adequate.

cont.

 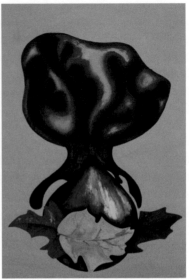

FIGURE 16.5 (a) Van Gogh's *Sunflowers* seen through a climate change lens. *Source*: Liz Stokoe. (b) *Droplet of Life* – an interpretation of water and climate. *Source*: Elizabeth Naden.

In the second part, a single image is displayed and the group must now invent a headline that fits two or more frames. For example, Figure 16.3 could support the following captions (and frames): 'Dawn of the age of the super-flood' (catastrophe frame); 'Business as usual despite record floods' (economic frame); or 'Scientists told us that the climate will get hotter and drier' (uncertainty frame). If time permits, the task could be repeated for other striking images (such as Figures 7.7 or 9.1).

LEARNING OUTCOMES

- To understand the major strengths and weaknesses of the 'deficiency' and 'cultural circuits' models of science communication.
- To recognise that climate messages are framed in different ways, depending on the target audiences, ideology and intentions of the information provider.
- To appreciate that effective communication (leading to increased scientific literacy) does not necessarily translate into more pro-climate behaviour.

16.3 FURTHER QUESTIONS TO RESEARCH AND DISCUSS

What are the most significant challenges faced by climate science
communicators who engage with public audiences in China AND
the United States?

How might climate change be framed for pre-teen school children?

In what ways has the rise of social media shaped public attitudes to
climate change?

16.4 FURTHER READING

Larson, K.L., White, D.D., Gober, P., Harlan, S. and Wutich, A. **2009**.
Divergent perspectives in water resources sustainability in a
public–policy–science context. *Environmental Science and Policy*, **12**,
1012–1023.

Moser, S. 2016. Reflections on climate change communication
research and practice in the second decade of the 21st century:
what more is there to say? *Wiley Interdisciplinary Reviews: Climate
Change*, **7**, 345–369.

16.5 OTHER RESOURCES

Climate Change Art https://uk.pinterest.com/mikaidt/climate-
change-art/ [accessed 16/07/16]

Climate Games www.climateinteractive.org/policy-exercises-and-
serious-games/19-climate-games-that-could-change-the-future/
[accessed 16/07/16]

Conker Tree Science Project www.conkertreescience.org.uk/
[accessed 16/07/16]

Media Coverage of Climate Change http://sciencepolicy.colorado.edu/
media_coverage/ [accessed 16/07/16]

Old Weather www.oldweather.org/#/ [accessed 16/07/16]

The Kids Carbon Calculator http://tfl.kopymark.com/ [accessed 16/07/16]

Tipping Point www.tippingpoint.org.uk/ [accessed 16/07/16]

Tomnod searches of satellite imagery for objects of interest to
science and policy www.tomnod.com/ [accessed 16/07/16]

UCAR Classroom Activities http://scied.ucar.edu/activities [accessed
16/07/16]

17

Who Are Climate Experts?

TOPIC SUMMARY

Climate expertise and scientific consensus shape public attitudes about global warming and attendant need for action. Expert opinion is also needed to define highly uncertain but critical variables (such as the climate sensitivity to CO_2 or likelihood of ice-sheet collapse). But measurement of expertise and consensus amongst experts is problematic. Although climate expertise and climate scepticism are not mutually exclusive, bibliographic analyses suggest that these groups have different publication profiles and affiliations. There are varied reasons for climate scepticism, including political bias and lack of certainty in scientific understanding. Nonetheless, positive contributions have been made through pushes for greater transparency and candour by climate scientists about key uncertainties. With the rapid emergence of a climate service community, there are now calls for more stringent, professional standards of practice and product. These ethical principles are needed to safeguard clients from possible maladaptation to climate change.

BACKGROUND READING

Morgan and Keith (1995) examine the role of expert judgement in quantifying uncertain elements of the climate system, and whether scientific consensus statements may underplay the true level of disagreement. Oreskes (2004) evaluates evidence of explicit or implicit acceptance of the consensus view in refereed journals.

17.1 WEIGHING EXPERTISE
War of the Words

A climate scientist and a climate change denier walk into a bar. The denier says, "*Bartender, show me your strongest whiskey*". The bartender says, "*This one here. It's 95 percent alcohol*". The denier slams down his fist and leaves the bar in a hurry. The scientist says, "*You know, that's the problem with these guys. You show them the proof, and they still don't buy it.*"[1]

What do Pope Francis, Leonardo DiCaprio and Homer Simpson have in common? This one is not a joke – they all have an opinion about climate change. Moreover, their views reach global audiences numbering millions, so what they say can respectively chastise deniers, motivate action or cast doubt on climate change. Does their influence mean that they are effectively climate experts? How much weight should be attached to the opinions of contrarian or authoritative but non-expert perspectives on climate change? Finally, despite the difficulty of measuring climate expertise, why does the label 'climate expert' matter so much?

This chapter explores the links between scientific expertise and the geopolitical imperative for urgent action on climate change. Some (quasi-)objective yet controversial methods of measuring expert credibility are introduced. These formal criteria are set alongside those used to weigh expert opinion or to represent divergent views on climate change in the media. The meaning and importance of expert consensus is then examined in relation to public acceptance of climate science messages. The contributions made by dissenting voices to climate science and policy are duly considered too. The final part shows how notions of expertise and ethics are increasingly pertinent to the fledgling climate service industry. Here, there is a danger that competing interests and motivations could have long-lived and far-reaching negative consequences for climate vulnerable communities.

Measuring Scientific Expertise

Type 'who are climate experts' into a search engine and close to the top of the 7 million plus results is a 'List of climate scientists' provided by Wikipedia. Dig a little deeper into the Wikipedia content

---

[1] www.die-klimaschutz-baustelle.de/new_global_warming_jokes.html

and it soon becomes apparent that the neutrality of the article can be disputed. In fact, the list was nominated for deletion in 2009 and the talk page raises concerns about the criteria for inclusion (apparently registering only scientists who have their own Wiki pages). Other notable scientists are excluded on the grounds that they are no longer 'active' (not least Svante Arrhenius)! There are also concerns that some 'run-of-the-mill' climate scientists may be listed, high-lighting the difficulty of gauging the significance of contributions to such a multi-disciplinary field.

There are many ways of defining expertise (see Collins and Evans, 2002). One debatable method is to count the number of authored or co-authored peer-reviewed publications and citations in relevant fields. For example, Anderegg et al. (2010) compiled a list of 1372 climate researchers who had contributed to the IPCC Fourth Assessment Report Working Group I, or who had signed major statements either endorsing (e.g. 2007 Bali Declaration) or dissenting (e.g. 2007 letter to the UN Secretary General Ban Ki-moon) the views of the IPCC. The list was checked for duplicate names and included only authors with at least 20 climate publications. More controversially, individual researchers were classified as either convinced (CE) or unconvinced (UE) of the evidence for anthropogenic climate change. The results showed that the UE group comprised only 2% of the top 50 researchers ranked by number of publications, and that the median number of their papers and citations was significantly lower than for the CE group. Bodenstein (2010) challenged the findings on the grounds that CE researchers were bound to achieve higher pub-lication metrics because of mutual citation within the mainstream group. He also objected to the emphasis on the *worth* of the climate change critics rather than on the *validity* of their arguments – echoing the common complaint that the number of citations does not equate necessarily with the merit of a research paper.

Even so, there appear to be measurable differences in the choice and use of scientific publications between mainstream and contrarian groups. Janko et al. (2014) compared the reference list of the IPCC Fourth Assessment Report Working Group I with that of the Heartland Institute's *Climate Change Reconsidered*. They found that the two groups often drew from the same journals for key sources; but the contrarians tended to cite more palaeoclimate science. Over-all, however, only 4.4% of the publications overlapped and, even then, the associated rhetoric was markedly different. Carlton and Jacobson (2016) assert that points of overlap between expert and non-expert (but not necessarily contrarian) 'mental models' of climate change provide entry points for more effective communication and outreach.

Trusting the Views of Experts

Sometimes scientific evidence does not exist or is simply too uncertain to support definitive statements about risks. Under such circumstances, climate scientists might offer their expert opinion or a probabilistic judgement. Examples of ambiguous knowledge include the future size of polar bear populations (O'Neill et al., 2008), carbon emissions from permafrost (Schuur et al., 2013), or pace of ice sheet melt (Bamber and Aspinall, 2013). Expert elicitations capture the known unknowns in climate models and assume that subjective probabilities can be attached to the uncertain quantities (Millner et al., 2013). For example, probability distributions of future West Antarctic Ice Sheet collapse and significant sea level rise can be built from the collective judgement of multiple experts based on their knowledge of the underlying risk factors (Figure 17.1).

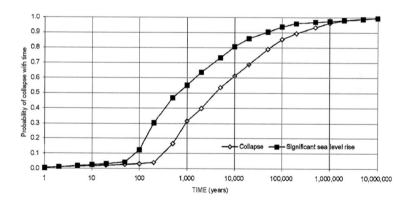

FIGURE 17.1 Estimates of the probability of collapse of the West Antarctic Ice Sheet (WAIS) and significant sea level rise (SLR), within a given time frame, based on the combined assessment of 11 experts. Here 'collapse' is defined as a change that would contribute at least 1 m of SLR per century, or 4 m in total. 'Significant sea level rise' is defined as at least 0.2 m per century, or 1 m in total. The graph shows that the experts regarded collapse as unlikely over the next few centuries but still retained a 5% probability of 2 m SLR or 30% probability of 0.4 SLR from WAIS melt over the next two centuries. *Source*: Vaughan and Spouge (2002)

Disputes can arise if the sample of experts or elicitation process is regarded as unrepresentative or *ad hoc*. In other words, ask a different set of experts and a different result can be obtained (see, for example, Lewis and Crok (2014) for an alternative view on climate sensitivity). Some have applied formalised approaches to more systematically weight and combine qualitative expert judgement with quantitative modelling (e.g. Abdallah et al., 2014; Oppenheimer et al., 2016). Others use hierarchical methods to rank and weight expert opinions on various climate policies (e.g. Berrittella et al., 2008). Alternatively, the Delphi technique can elicit expert opinion on high-risk, low-probability hazards (e.g. Vaughan and Spouge, 2002) or define successful adaptation to climate change (e.g. Doria et al., 2009). This system involves panels of experts with knowledge of the area completing structured questionnaires and then iteratively reviewing group-wide opinions.

Expertise is framed differently by the mass media. Boykoff (2013) complains that some parts 'gratuitously' over-represent contrarian views on climate change because of journalistic norms, institutional practices and external political factors. In seeking a balanced debate of the science, some media actually convey a narrative that is biased disproportionately in favour of the contrarian minority (Boykoff and Boykoff, 2004). This narrows the possibilities for action and conveys a polarised perspective on climate expertise. On the other hand, those who pay close attention to TV weather forecasts are more likely to believe that extreme weather is becoming more frequent and to be more concerned about climate change (Bloodhart et al., 2015). This suggests that a trusted weather forecaster can, in effect, be the local expert to some public audiences. Teachers are also instrumental in (mis-)shaping attitudes, so their views matter. A survey of 1500 public middle- and high-school science teachers in the United States found that 30% emphasise natural causes as the reason for recent global warming, with 2% still denying that there is any warming (Plutzer et al., 2016).

Consensus and Public Perceptions of Climate Change

One analysis of peer-reviewed scientific papers with abstracts expressing a position on anthropogenic climate change found that 97% endorsed the view that humans are causing global warming (Cook et al., 2013). The statistic was further corroborated through self-rating of papers by the authors. However, with the aforesaid balanced (i.e. 50:50) reporting of climate change by the media, it is hardly surprising that the public underestimate the extent of consensus amongst climate scientists. One study of 139 respondents found that the public perceived 70% scientific consensus (Lewandowsky et al., 2013). This mismatch is important because the same study showed that public acceptance of anthropogenic climate change increases when consensus is highlighted. Other experiments confirm that scientific consensus is a 'gateway belief' – a causal factor that ultimately influences the level of public support for public action on climate change (van der Linden et al., 2015).

Unfortunately, the case for action built around consensus can also be an edifice that is more readily shaken. This was evident in the aftermath of the alleged misconduct of a few climate scientists during the 'Climategate' affair (see below) (Lahsen, 2013). The scientists were cleared of wrongdoing but the inquiries did raise questions about the transparency and rigour of some temperature analyses. This is important because debates about climate science and climate policy are so intimately tied that pointing to uncertainty in the science and/or creating distrust in the scientist are effective ways of undermining public faith in climate policies (Dahan-Dalmedico, 2008; Sarewitz, 2011). Hence, the 97% consensus statistic has become a battleground for opponents of action on climate change (e.g. Idso et al., 2015). Principle objections include that (i) the articles of prominent climate change sceptics are often overlooked in bibliographic assessments of consensus; (ii) surveys do not indicate whether the respondents see anthropogenic global warming as dangerous; and (iii) the methodological design and execution of surveys are biased in favour of the non-contrarian view.

Causes and Value Added by Climate Scepticism

The roots of climate scepticism are manifold and complex. Of course, there are those individuals, corporations and nations whose financial interests are opposed to cuts in GHG emissions. According to Jacques (2012), acceptance of the orthodoxy of climate science carries with it a realisation that the existing global economic and political order is unsustainable. Much environmental scepticism can be traced to conservative think tanks (Jacques et al., 2008). Indeed, political ideology is one of the strongest predictors of US attitudes to global warming, as evidenced by surveys conducted by the Pew Research Center. These consistently show wide partisan divides. For example, one poll taken in 2015 revealed that 83% of conservative and moderate Democrats believe that the Earth is warming compared with just 38% of conservative Republicans.[2]

Capstick and Pidgeon (2014) reveal some more nuanced aspects of climate change scepticism. They make a distinction between epistemic (scientific) and responsive (actions) scepticism. Variations in public scepticism of scientific authority also occur between and within countries. In Germany, the reality of anthropogenic climate change is almost universally accepted. Nonetheless, the political and democratic legitimacy of the IPCC in shaping domestic climate policies cannot be taken for granted (Beck, 2012). In New Zealand, psychological acceptance of climate change increases with proximity to the coast, presumably due to greater perception of risk from flooding, storms and rising sea levels within seaside communities (Milfont et al., 2014). Likewise in the United States, for some adults, direct experiential learning from extreme weather events may be more convincing than scientific arguments (Myers et al., 2013). In the UK, scepticism about global warming is not particularly firmly held or widespread; doubt is mainly found amongst those who are 55 and over, from lower socio-economic groups, and/or likely to vote Conservative (Poortinga et al., 2011).

Legacy of Climategate

Much has been written about the negative tactics of misinformation, manufacture of doubt and organisation of science denial (e.g. McCright and Dunlap, 2003; Jacques et al., 2008; Lewandowsky et al., 2012; Rosenau, 2012; Boykoff, 2013). Climategate provides an example of how mixed outcomes emerged from what was an illegal act, motivated by climate scepticism. The hacking and publication of e-mail exchanges between leading climate scientists occurred just before the Copenhagen COP meeting in 2009. According to Maibach et al. (2011) TV weather presenters who followed the story reported less trust in the IPCC and climate scientists than those who had not. Discourse analysis of readers' comments in UK tabloids before and after the affair revealed increased use of derogatory language in relation to the term 'scientist' (Koteyko et al., 2013). Nationally representative surveys of US public attitudes showed that the scandal had a significant impact on climate

---

[2] www.pewresearch.org/fact-tank/2015/06/16/ideological-divide-over-global-warming-as-wide-as-ever/

change beliefs, risk perceptions and trust in scientists (Leiserowitz et al., 2013). [But see as light relief associated climate cartoons, e.g. www.cartoonsbyjosh.com/blurb.html.]

On the other hand, Grundmann (2012) asserts that the affair prompted wider calls for greater transparency and openness in the practices of climate scientists. The ethical conduct and nature of climate science–policy interactions have also since come under closer scrutiny (Grundmann, 2013). Journals and research funders now routinely require that underpinning data are open access, and that authors declare any competing financial interests. More importantly, there has been greater attention to the transparency and procedures of the IPCC; including the value of meta-data and quality assurance of long-term climate records. A further outcome has been more fruitful scientific discussions about climate model limitations and their ability to simulate recent multi-decadal variability (Table 17.1). Although some see debate of the putative global warming hiatus (apparent slowdown in warming between 1998 and 2014) as 'seepage' of climate change denial into the scientific community (Lewandowsky et al., 2015), others regard the mismatch between climate models and recent temperature trends as a legitimate subject for further research (Curry, 2014a).

Ethical Climate Services

Healthy scepticism has always been important to the development of science. Conversely, downplaying scientific uncertainty or over-stating confidence is disingenuous, even dangerous in some situations. Thankfully, cases of scientific fraud are rare and increasingly detectable using linguistic tools (e.g. Markowitz and Hancock, 2015). Nonetheless, hyperbole is on the rise in climate research papers (Figure 17.2). Between 1991 and 2015, the occurrence of positive words (such as 'groundbreaking') in the titles or abstracts of 'climate change' papers more than doubled. Whether climate change sceptic or mainstream climate scientist, we would all benefit from a dose of humility when it comes to recognising our own knowledge limitations (Ferkany, 2015). This means presenting climate science as no more or less than it is.

Humility sits alongside integrity, transparency and collaboration as core values that should be integral to an ethical framework for climate services (Adams et al., 2015). The emerging climate service industry provides tools, products and information to help clients anticipate and address both threats and opportunities from climate change. The four values were used by the Climate Services

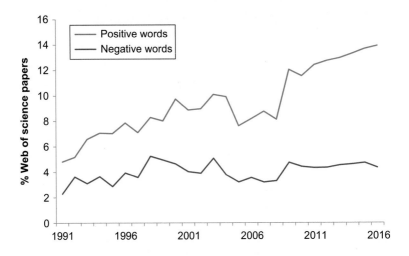

FIGURE 17.2 The percentage of papers with positive and negative words used in the title, abstract or keywords of 'climate change' or 'global warming' papers listed on Web of Science. Following Vinkers et al. (2015) the 25 positive words were any of: amazing, assuring, astonishing, bright, creative, encouraging, enormous, excellent, favourable, groundbreaking, hopeful, innovative, inspiring, inventive, novel, phenomenal, prominent, promising, reassuring, remarkable, robust, spectacular, supportive, unique, unprecedented. The 25 negative words were any of: detrimental, disappointing, disturbing, frustrating, futile, hopeless, impossible, inadequate, ineffective, insignificant, insufficient, irrelevant, mediocre, pessimistic, substandard, unacceptable, unpromising, unsatisfactory, unsatisfying, useless, weak, worrisome.

Partnership to lay down a set of working principles of practice and product for climate service providers (Table 17.2). Their intention is to contribute to human security at individual and community levels, thus placing a strong onus on avoiding poor decision-making and maladaptation. Monetising the benefits of climate services is not straightforward, not least because many intended adaptation benefits may be in the future (e.g. lives and property saved by constructing a flood defence that makes allowance for changing river flows under climate change). More generally, there is a lack of measurable adaptation outcomes (Vaughan and Dessai, 2014) and concerns that climate services might provide uneven benefits (Pfaff et al., 1999). For instance, not every smallholder has the resources to benefit from seasonal forecasting by adjusting the time of planting, choice of crop or fertiliser application (see Chapter 8).

TABLE 17.1 Battle of the blogosphere: examples of sites that promote different perspectives on anthropogenic climate change [all accessed 13/6/2016]

Perspective	Name	Link
Consensus	Carbon Brief	www.carbonbrief.org/
	Climate Central	www.climatecentral.org/
	Real Climate	www.realclimate.org/
Environmental	Friends of the Earth	www.foe.co.uk/campaigns/climate_change
	The Nature Conservancy	www.nature.org/ourinitiatives/urgentissues/global-warming-climate-change/
	WWF	wwf.panda.org/what_we_do/footprint/climate_carbon_energy/
Faith-based	Christian Aid	www.christianaid.org.uk/whatwedo/issues/climate_change.aspx
	Jewish Climate Action Network	www.jewishclimate.org/#home
	Muslim Climate Action	www.muslimclimateaction.org.uk/
Humanitarian	Oxfam	www.oxfam.org.uk/what-we-do/issues-we-work-on/climate-change
	Red Cross Red Crescent	www.climatecentre.org/
	Save the Children	www.savethechildren.org/site/c.8rKLIXMGIpI4E/b.9353925/k.80C2/Climate_Change.htm
Luke warmers*	Climate Etc.	https://judithcurry.com/
	Climate Lab Book	www.climate-lab-book.ac.uk/
	Matt Ridley	www.rationaloptimist.com/blog

cont.

TABLE 17.1 (cont.)		
Perspective	Name	Link
Opposed to action	Global Warming Policy Foundation	www.thegwpf.com/
	Heartland Institute	www.heartland.org/
	Watts Up With That?	wattsupwiththat.com/

* This is a non-pejorative term for those who seek to uncover and resolve outstanding knowledge gaps in climate science

TABLE 17.2 Principles of practice and product for ethical climate services. Source: Adams et al. (2015)	
Principles of practice **Climate service providers should:**	*Principles of product* **Climate service products should:**
(1) communicate value judgements;	(1) be credible and defensible;
(2) communicate principles of practice;	(2) include detailed descriptions of uncertainty;
(3) engage with their own community of practice;	(3) be fit for purpose; and
(4) engage in the co-exploration of knowledge;	(4) be documented.
(5) understand climate as an additional stressor;	
(6) provide metrics of the skill of their products;	
(7) communicate appropriately;	
(8) articulate processes for refreshing and revising their products and information;	
(9) have mechanisms for monitoring and evaluation of procedures and products; and	
(10) declare any conflicts of interest and/or vested interests.	

Strengthening the climate services of less-developed countries is an essential part of protecting development gains from climate variability and change (Jones et al., 2015). Given the high climate vulnerability of developing regions (Figure 7.3) the opinions of climate experts can matter a great deal. Humility enters the equation once again, in terms of greater emphasis on listening to user needs before talking about what the science can provide. It is also important to understand the contextual barriers to information uptake. These might include the *spatial and temporal scale, accessibility, timing, credibility and the mismatch in timeframes between planning cycles (1–5 years) and climate projections (over 20 years)* (Vincent et al., 2015:1). Above all, climate service providers have to be mindful of the consequences of their actions and uphold professional standards that go beyond the norms of scientific practice.

So You Think You Are an Expert?

Well done. You are nearly at the end of this book, but does that make you a climate change expert? What values would you cite for climate sensitivity or for the probability of polar bear extinction by the end of this century? How much confidence (%) would you attach to each statement, and what knowledge or experience sits behind your judgement? Are you prepared to accept the responsibility for major investments or for your work to be subject to unprecedented scrutiny? Would you strive to meet the high ethical standards being set for climate service providers?

To an extent, and largely thanks to social media, we are now all experts in the era of 'crowd sourcing' and 'citizen science'. However, climate expertise really lies in the collective knowledge and experience of a broad community of practice, rather than within the narrow domains of a few prominent researchers. The final chapter situates climate change within a diverse set of concerns about the future – a further reminder that an open and inclusive perspective is needed to meet the climatic and non-climatic challenges ahead.

17.2 GROUP EXERCISE: ETHICAL DILEMMAS OF A CLIMATE SERVICE PROVIDER

NOTES

This activity touches on some ethical dilemmas that might be encountered at various stages in a career, or within different parts of the climate service industry. It may be helpful to begin by determining whether the scenario invokes the principles of *practice* or principles of *product*, or both (see Adams et al., 2015).

Depending on the size of the group it may be more effective to allocate scenarios to sub-clusters then collate responses, rather than have the entire class work through all scenarios. A summary of the Climate Services Partnership principles (Table 17.2) should be distributed or displayed before the start of the exercise.

Ask the group to imagine that they are a climate service provider (i.e. consultant, researcher, government employee) and discuss how they would handle the following situations:

Scenario 1: You are in the last few weeks of a 12-month consultancy project about major droughts and discover a fundamental flaw in one of your early analyses. This mistake is very hard to spot and, even if fixed, would not materially affect the headline findings of the study (as far as you can tell). What are the ethical issues? What would you do?

Scenario 2: You work for a company that sells ultra-high-resolution (<5 km scale) scenarios of heavy rainfall for the 2050s and 2080s. A potential customer in a rapidly growing city is interested in buying the information to design a flood embankment to protect 250,000 people in a wealthy district. The structure would lead to the eviction of others. What are the ethical issues? What would you do?

Scenario 3: You are an early career climate scientist and supplement your grant with modest consultancy income. You complete a small project that involved some innovative modelling. A decade later you discover that the client published your work under his own name with no acknowledgements to your contribution. What are the ethical issues? What would you do?

cont.

Scenario 4: You are hired by international investors to evaluate the feasibility of a major hydropower scheme. Your results are very different from the findings of another consultancy firm. The client asks you to reconcile your findings with the other group. What are the ethical issues? What would you do?

Scenario 5: You are asked by a small renewable energy company to analyse data about one of their new technologies. They want to demonstrate high energy yields to potential investors in their product. Even a quick look at the data tells you that high yields are unlikely. What are the ethical issues? What would you do?

Scenario 6: You are proofreading a speech to be given by your boss at a very high-profile debate about funding of coastal flood defences for vulnerable communities. You feel that some statements about future coastal flood risk and erosion are exaggerated. What are the ethical issues? What would you do?

LEARNING OUTCOMES

- To understand the various methods and challenges of measuring scientific expertise and consensus, plus why these matter to public opinions about climate change.
- To be aware of the many factors that affect acceptance of global warming, including political ideology, socio-economic status and even geographic location.
- To recognise the importance of humility, integrity, transparency and collaborative working as the bedrock for ethical climate services.

17.3 QUESTIONS TO RESEARCH AND DISCUSS

There is scientific consensus about anthropogenic global warming: matter of fact or myth?

Give specific examples of the ways in which the conduct and integrity of climate science have been strengthened by the activities of climate sceptics.

To what extent do the procedures and reports of the IPCC achieve the ethical standards of practice and of product laid out by Adams et al. (2015)?

17.4 FURTHER READING

Bray, D. 2010. The scientific consensus of climate change revisited. *Environmental Science and Policy*, **13**, 340–350.

Dunlap, R.E. and McCright, A.M. 2010. Climate change denial: sources, actors and strategies. In Lever-Tracy, C. (ed.), *Routledge Handbook of Climate Change and Society*. Routledge, Abingdon.

McCright, A.M., Charters, M., Dentzman, K. and Dietz, T. 2016. Examining the effectiveness of climate change frames in the face of a climate change denial counter-frame. *Topics in Cognitive Science*, **8**, 76–97.

17.5 OTHER RESOURCES

European Commission Road Map for Climate Services www.gppq.fct.pt/h2020/_docs/brochuras/env/A_European_research_and_innovation_Roadmap_for_Climate_Services.pdf [accessed 18/07/16]

Global Framework for Climate Services http://gfcs.wmo.int/ [accessed 18/07/16]

Statement by the American Meteorological Society on climate services www2.ametsoc.org/ams/index.cfm/about-ams/ams-statements/statements-of-the-ams-in-force/climate-services1/ [accessed 18/07/16]

Survey by the American Meteorological Society on views about global warming http://journals.ametsoc.org/doi/abs/10.1175/BAMS-D-13-00091.1 [accessed 18/07/16]

18

How Connected Is Climate Change to Other Global Challenges?

TOPIC SUMMARY

It is tempting to think that climate change is the most important global challenge facing humanity. Perhaps it really is. However, climate change should be seen as connected to, rather than apart from, other concerns about shelter, water, food and energy security. Some fear that these pressures could converge in a 'perfect storm'. The nexus concept acknowledges interdependencies and points to multiple wins such as improved water efficiency of crops, requiring less fuel for pumping groundwater, leading to lower GHG emissions. The climate nexus is discussed for three sets of related concerns: (i) water–food–energy; (ii) biodiversity–food–energy; and (iii) livelihood–migration–conflict. These show that climate shocks are rarely the only or most important pressure on human and natural systems. Scope for addressing these challenges is improving thanks to unprecedented monitoring of Earth systems, a more pragmatic approach to curbing emissions, and growing capacity to deploy climate services when and where they are most needed.

BACKGROUND READING

Liu et al. (2015b) call for more integrated, cross-sectoral approaches to global challenges such as pollution, biodiversity loss, climate change, water scarcity, energy and food security, spread of disease and invasive species. Rasul and Sharma (2015) explain how the nexus concept helps to identify trade-offs and strengthen the security of water, food and energy resources.

18.1 AVOIDING CLIMATE EXCEPTIONALISM
Worlds of Worry

According to a Pew Research Center poll of public policy priorities for 2015, the top three concerns of US citizens were terrorism, the state of the economy and jobs.[1] Global warming was ranked 22nd – just above global trade. But recent years have seen growth in the percentage of respondents saying that dealing with the issue should be a priority for the Administration. Overall, greatest public concern about climate change was found in parts of South America, sub-Saharan Africa and South Asia. Citizens of Brazil, Burkina Faso and Peru were most worried, whereas those surveyed in Australia, Israel and the United States were least concerned.[2]

It seems that people in countries with the highest carbon emissions per capita were generally less worried about climate change and potential impacts. Across all regions, the climate threat that causes greatest anxiety is drought, followed by severe weather, heat-waves and rising sea levels. When asked in 2014 what are the greatest threats to the world,[3] respondents from 44 countries top-ranked religious and ethnic hatred (24%) with inequality (24%), ahead of nuclear weapons (22%), pollution and environment (13%), AIDS and infectious diseases (10%).

Responsible governments do not have the luxury of focusing on a single concern. As resources permit, they must try to address them all. Unfortunately, the above 'to do list' is far from complete. There are other fears about (i) food, water and energy security; (ii) loss of biodiversity and spread of invasive species; (iii) regional conflict and mass migration. Beddington (2009) famously spoke of a 'perfect storm' of food, water and energy shortages by 2030. The former UK Government Chief Scientific Advisor warned that this tempest would have to be faced whilst at the same time contending with population growth, mounting competition for scarce resources and climate change. In his vision of the future, the myriad global challenges are not separate but intrinsically linked.

This chapter traces the presence of climate change amongst the three sets of concerns mentioned above (i.e. resource security,

---

[1] www.people-press.org/2015/01/15/publics-policy-priorities-reflect-changing-conditions-at-home-and-abroad

[2] www.pewresearch.org/fact-tank/2016/04/18/what-the-world-thinks-about-climate-change-in-7-charts/

[3] www.pewglobal.org/2014/10/16/greatest-dangers-in-the-world/

species richness and regional conflict). The term 'nexus' is used to stress the *connectivity* of climate with natural and human systems. There are other important influences or so-called 'megatrends' (Table 18.1) – such as demographic change as a driver of energy demand. So the intention is to close with a calibrated view of the risks posed by climate change and counter what some say is 'climate exceptionalism' (Nagle, 2009; Curry, 2014b; *The Economist*, 2014). That is, to think that climate change is a uniquely wicked problem (which it is not). Or to blame climate change for everything that is wrong in the world (which it is not). The final section steps back from all this sombre talk to give three reasons why there is still room for cautious optimism.

The Climate–Water–Food–Energy Nexus

The United Nations Food and Agriculture Organization (FAO) estimate that the amount of precipitation falling on land each year is ~110,000 km^3. About 56% of this water is evapotranspired from natural landscapes, 39% enters rivers and aquifers, and the remaining 5% evaporates from rain-fed agriculture. Of all the withdrawals from surface and groundwater sources – so called 'blue' or renewable water – 69% is taken for agriculture (irrigation, livestock and aquaculture), 19% for commercial and 12% for municipal uses. In 2010, global withdraws (including water evaporated from artificial lakes) summed to ~4250 km^3 out of a total renewable resource of 43,000 km^3. At face value there appears to be plenty of spare water for food and energy production. However, closer analysis reveals that this assumption is wrong.

First, global statistics conceal large regional variations in water availability for irrigated agriculture (Figure 18.1). Data from UN AQUASTAT suggest that North Africa is presently using 77% of renewable freshwater resources for irrigation; in the Arabian Peninsula it is 471% of replenished supply. By comparison, Western Europe and North America use only 0.5% and 3.7% respectively. In 2006, Saudi Arabia's water withdrawals were a staggering 943% of supply. Second, water demand is out-stripping the growth rate of population – during the twentieth century population increased by a factor of 4.4, whereas water withdrawals rose by a factor of 7.3. Rockström et al. (2007) estimated that in order to alleviate hunger amidst a global population of 9 billion, another 5200 km^3 of water will be needed for agriculture by 2050. Some of this water demand could be met by efficiency gains and/or

TABLE 18.1 Megatrends that are faced by governments as they discharge their core responsibilities for national security, social cohesion, economic prosperity and environmental sustainability. Adapted from KPMG (2014).

Megatrend	Description
Demographics	Higher life expectancy and falling birth rates lead to an ageing population with challenges for social welfare systems in some regions, or higher life expectancy and high birth rates leading to youth unemployment in other regions
Rise of the individual	Improved education, health and technology are empowering individuals whilst increasing demands for transparency and accountability of governments and public decision-making
Enabling technology	Information and communications technology is transforming society, changing patterns of employment and demand for skills and testing regulatory powers
Economic interconnectedness	Globalisation of international trade and flow of capital has the potential to increase economic benefits but also harm from global downturns
Public debt	Debt is a significant constraint on fiscal and policy options for delivering public services by governments or responding to emergent social, economic and environmental challenges
Economic power shift	Emerging economies are lifting millions out of poverty and changing the balance of global power whilst international institutions and foreign ownership transcend state authority

cont.

TABLE 18.1 (cont.)	
Megatrend	Description
Climate change	Rising concentrations of GHGs change the climate and drive a range of impacts on natural and human systems with costs and benefits to society
Resources stress	Combined effects of population growth, economic growth and climate change place sustainable water, food, energy and land resources under increasing stress
Urbanisation	Two-thirds of the global population could be living in built areas by 2030, creating opportunities for socio-economic development and more sustainable life-styles but concentrating resource demand and vulnerability to hazards

higher productivity (i.e. more crop per drop) but they estimated that ~40% would come from expanding rain-fed agriculture into areas presently occupied by tropical forests and grasslands.

Water scarcity is one amongst many drivers of global food (in)security. Foresight Futures (2011) and Hanjra and Qureshi (2010) cite many factors including population growth, changing diets (such as more meat consumption), energy prices, development of new crop and livestock varieties, use of land for ecosystem services or economic activity, level of investment in agriculture, globalisation and the international food trade, increasing urbanisation, robustness of supply chains, global security, amount of food waste, income distribution, human health, technological innovation AND climate change. As discussed before, more extreme and changing regional patterns of flood and drought are expected to significantly impact future crop production (Figures 11.2 and 11.3). In the Hindu Kush Himalayan region, there is already a high degree of dependency of downstream communities on upstream ecosystem services for dry-season water for irrigation and hydropower, drinking water, soil fertility and

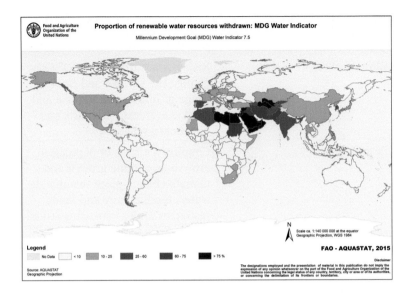

FIGURE 18.1 Proportion of renewable water resources withdrawn (%).
Source: Food and Agriculture Organization of the United Nations (2016)
AQUASTAT, www.fao.org/nr/water/aquastat/maps/index.stm. Reproduced
with permission.

nutrients (Rasul, 2014:35). In situations like this, management of
changing transboundary water supplies adds further challenges.

Improved efficiency of groundwater use may yield both water and
energy savings. Energy expended lifting groundwater for irrigated
agriculture has grown dramatically over the last three decades and,
for example, now accounts for 0.5% and 3.6% of total national
carbon emissions of China (Wang et al., 2012) and Iran (Karim
et al., 2012), respectively. Simply adjusting irrigation schedules can
improve application efficiencies, whilst maintaining productivity at
the farm scale. Energy consumption could be reduced by as much as
40%, depending on how much of the water saving is not redeployed
elsewhere (Karim et al., 2012). More controversially, policies that
raise tariffs for agricultural energy use could retard aquifer deple-
tion by incentivising water saving (Scott, 2011).

Other major water users could also realise substantial savings. For
instance, it has been estimated that water cooling of thermoelectric
power plants in Texas consumes the equivalent annual water used by
3 million people. Power generation efficiency is less for air- than
water-cooled plants, but the water saving for a coal-fired plant could
be ~99.7% (Stillwell et al., 2011). More radical suggestions include

using electrical grids to shift power production (with water cooling) away from areas stricken by drought (Pacsi et al., 2013). This expensive option would be quicker to implement than construction of new water pipe-lines or retrofitting with air cooling technologies. In the UK, there may be a long-term case to shift generating capacity away from rivers to estuaries or the coast to adapt to expected reductions in low flows under climate change (Byers et al., 2015). A different mix of technologies is needed to achieve both water saving and reductions in CO_2 emissions than would otherwise be deployed to reduce one or the other (Webster et al., 2013).

National assessments of water resources do not tell the full story because each country donates and receives water embodied in commodities. Figure 18.2 shows the international flow of virtual water that has been used in the production of wheat. For example, the United States exports the equivalent of 7 km^3 (Gm^3) of virtual water per year to Japan alone. On the other hand, Brazil receives more than 10 km^3 of water per year embodied in wheat imports from Argentina. The map is for a single commodity – now imagine the amount of virtual water flow linked to international trade. For the UK alone, imports of rice, meat, plastics, paper and other water-intensive products effectively transfer 13 km^3 of water per year from exporting countries (Hunt et al., 2014). With ~70% import-water dependency, the UK is potentially vulnerable to climate disruption of supply chains (Chapagain and Hoekstra, 2008). Although imports of commodities such as rice cannot be substituted by domestic production, water embodied in foreign goods such as dairy and meat

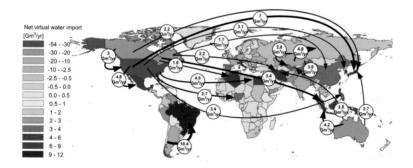

FIGURE 18.2 National virtual water balances and net virtual water flows related to trade in wheat products during the period 1996–2005. Only the largest net flows (>2 Gm³/yr) are) shown. *Source*: Mekonnen and Hoekstra (2010)

is effectively alleviating pressure on national freshwater environments (Hunt et al., 2014). Wealthy but water-stressed countries such
as Saudi Arabia have deliberately out-sourced food production to
safeguard their own scarce water resources for more economically
productive uses (DeNicola et al., 2015).

The Climate–Biodiversity–Food–Energy Nexus

More water and/or land for food and energy production may be taken
from natural ecosystems (Rockström et al., 2007). These habitats will
be simultaneously stressed by a host of other human pressures AND
climate change (Figure 18.3). Fundamental changes in community
composition are expected within nature reserves too (Thuiler et al.,
2006). Take, for example, the plight of populations of giraffe and
other savannah ungulates (three-toed large mammals), which have
declined at an alarming rate in recent decades. The International
Union for Conservation of Nature estimated that in 1991 there were
140,000 giraffes in the wild; now there are thought to be less than
80,000. The major threats to giraffe populations in North Africa are
from armed conflict, poaching and habitat degradation with or
without climate change. More generally, some ungulate populations
could be extirpated where boundary fencing restricts migratory
responses by large mammals to drought and rising regional

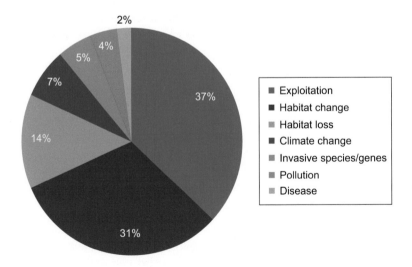

FIGURE 18.3 Primary threats to Living Planet Index (LPI) populations. *Data
source*: WWF Living Planet Report 2014

temperatures (Ogutu and Owen-Smith, 2003). The first mammal extinction linked to climate change was recently declared by the State of Queensland. Due to sea level rise and erosion, the area above high tide on Maizab Kaur Island in the Great Barrier Reef has shrunk to less than 2.5 ha and thereby eliminated the only known habitat of the small rodent Bramble Cay melomys (Gynther et al., 2016). This case highlights the extreme vulnerability of endemic species that are confined to small areas of climate-sensitive habitat.

Other habitats and food-producing lands are being lost or degraded as a consequence of policies to increase renewable (low carbon) energy production. According to the FAO *Statistical Year Book 2013*, less than 0.01% of the total global agricultural land (5 billion ha) is used for biofuel production, but this area was half the size in 2005. Globally, land use for rice, coconut, rubber and palm oil production in Indonesia, Malaysia and the Philippines is thought to contribute to most biodiversity loss (Chaudhary and Kastner, 2016). Moreover, clearing of rain forest, peatland, savannah or grassland for biofuel crops releases 17–420 times more CO_2 than would be displaced by use of the biofuel (Fargione et al., 2008). The International Energy Authority calculate that net carbon savings range from 10% for biofuels from maize production in the United States to 80% lower life-cycle emissions from sugar cane production in Brazil. Less harmful biofuel feedstock can be produced on degraded lands abandoned for food cultivation, or extracted from crop residues such as wheat straw, waste from sustainably harvested forests, via double or mixed cropping, or from municipal waste streams (Tilman et al., 2009). Others assert that potentially negative effects of biofuel cultivation on biodiversity can be eased (but not eliminated) by careful site-specific crop selection (Eggers et al., 2009).

International commodity trading and travel have greatly facilitated the spread of invasive species. Climate change is expanding the bioclimatic envelope (range) within which these new arrivals might settle, and/or shrinking the habitable space for established biota. Mainka and Howard (2010) refer to a *double jeopardy* of climate change and invasive species acting together. Climate change potentially impacts species through modified behaviour, phenology (timing of life-stages), predator–prey relationships, habitat, disturbance regimes (such as fire and storms), as well as the spread of diseases, pests and competitor species (Figure 18.4). Early models estimated that climate change could 'commit' 18–35% of sampled taxa (species groups) to extinction by 2050 (Thomas et al., 2004).

Nevertheless, recent data syntheses show wide-ranging outcomes for *net* changes – positive and negative – in global species richness depending on different interpretations of spatially biased, locally representative and typically short-lived records (Gonzalez et al., 2016). Switchgrass and giant reed are examples of biofuel species that have the potential to become invasive (Barney and DiTomasco, 2008). Examples of invasive non-plant species that threaten agricultural productivity include the northward expansion of the genus *Culicoides* (biting midges that are the vector for the blue-tongue virus) (Purse et al., 2005; Samy and Peterson, 2016) or the fungal disease wheat leaf rust (Junk et al., 2016; Ziska et al., 2011).

The Climate–Livelihood–Migration–Conflict Nexus

There has been much speculation about possible links between climate shocks and (water, food, land) resource scarcity, vulnerable populations, mass migration and armed conflict (e.g. Gleick et al., 1994; Burke et al., 2009; McMichael, 2012). Such causation harks back to an era of environmental determinism during which it was thought that the geographical context can predispose societies to certain development pathways, or that climate change could lead to the collapse of past civilisations (e.g. drought and the Maya; Haug et al., 2003). Simple linear cause-and-effect thinking is rejected by most modern scholars of climate and conflict (e.g. Salehyan, 2008;

(a)

FIGURE 18.4 Modelled (a) current (1964–2014) and (b) future (2050s) ecological niche of blue-tongue virus under four RCPs. Shaded areas are modelled suitable conditions; white areas are unsuitable. The future expansion is greatest in central Africa, the United States and western Russia due to projected changes in areas with favourable bioclimatic conditions (temperature and precipitation). *Source*: Samy and Peterson (2016)

(b)

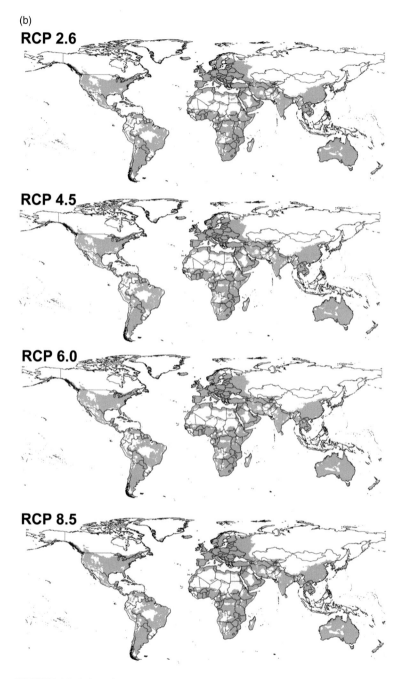

FIGURE 18.4 (*cont.*)

Buhaug, 2010; Kallis and Zografos, 2014). One criticism is that the future climate and socio-economic context will have no analogue in historical climate conflicts. There is also a need for firmer theoretical grounding to explain possible causality and to account for situations where there has been no conflict despite apparent climate risks (Gemenne et al., 2014). The prevailing view is that conflict emerges from a more nuanced and adaptive system of demographic, social, cultural, economic and political responses to environmental pressures. In such a systems-based framework, climate change is regarded as a threat multiplier (Buhaug et al., 2008) with various pathways to conflict (Burrows and Kinney, 2016).

For example, Benjaminsen et al. (2012) found little evidence linking erratic rainfall to violent farmer–herder disputes in the Mopti region of Mali. Instead, the key drivers of conflict were corruption among government officials, agricultural encroachment into the rangelands of herders, and political vacuum enabling opportunistic land grabs. Nonetheless, the Sahelian droughts of the 1970s and 1980s did contribute to conflict by reducing flood levels in the Niger delta and increasing encroachment of rice paddy fields into areas previously used to grow dry-season cattle fodder. Confrontations also occurred here in good rainfall years when the flood zone expanded into areas of ambiguous land ownership and control. The importance of land use change as the material link between climate and conflict has also been noted for North Africa (Link et al., 2015).

The climate–food–migration nexus integrates social vulnerability and geopolitical context with climate pressures to explain major episodes of environmental degradation, hardship and population movement (Piguet, 2010). For example, dependency on just two crops (potatoes and oats) and declining food entitlements afforded little resilience to the cold winter and subsequent Irish famine of 1740–1741 (Engler et al., 2013). In this case, emigration was interpreted as a form of adaptation. High-dietary reliance on a few crops is still regarded as problematic in the twenty-first century. For instance, the 2008–2010 global food crises highlighted the vulnerability of the Middle East to climate shocks in wheat supplies, West Africa to reduced supply of rice, and Central America to interruptions in maize supply (d'Amour et al., 2016). In these regions, adaptation responses might include increased agricultural productivity, diversification of diets and better supply chains. Surprisingly, migration was not mentioned in this study. In fact, surveys in Malawi suggest that acute food shortages erode human, financial and social capital to the point where migration aspirations and capabilities are actually reduced (Suckall et al., 2016). The amount of climate-related migration in sub-Saharan drylands also appears to depend strongly on population growth (Kniveton et al., 2012).

The climate–migration pathway mentioned above is cited as one route by which climate variability and change might spark conflict. The bloody civil war in Syria, which began in 2011, has been blamed partly on drought, climate change and migration. Gleick (2014) asserts that an extreme multi-year drought caused the deterioration of Syria's economy, increasing rural–urban migration and leading to high rates of unemployment and social unrest. It is estimated that the drought caused wheat and barley yields to fall by 47% and 67%, respectively, between 2006 and 2009, affecting 1.3 million inhabitants of eastern Syria. This all preceded the wave of turmoil surrounding the Arab Spring, which may have been triggered, in part, by climate-related global food crises in 2011 (Johnstone and Mazo, 2011). In contrast, document analysis and interviews with policymakers suggest that Israel and Jordon framed the subsequent international migration of Syrian refugees as a security issue driving increased pressure on their own scarce water resources; in Syria connections between drought, rural neglect and internal migration have been downplayed by officials (Weinthal et al., 2015).

Optimistic Outlook

It would be misleading to end with a litany of human suffering, all linked to climate. On this very day of writing (24 June 2016), the UK electorate voted to leave the EU. In a blink, the nation's geopolitical position looks very uncertain. Aside from some heavy downpours and travel disruption in London on referendum day, we cannot blame the 'leave' result on climate. But this single act could have far greater consequences for the socio-economy and environment than any conceivable extreme weather events or climate change over the coming decade. Time will tell whether the appetite of 'Little Britain' to mitigate and adapt to climate change has been set backwards.[4] One of the first actions of the reformed Cabinet was to shut down the Department of Energy and Climate Change. The carbon price has already slumped, and UK ratification of the Paris Agreement has been pushed

[4] Brexit and climate change https://judithcurry.com/2016/06/25/brexit-and-climate-change/

to the end of 2016. Beyond these immediate concerns there are still at least three reasons for cautious optimism.

First, the international community has unprecedented amounts of information and analytical tools with which to understand the evolving state of the planet. We have moved from a small number of stations monitoring air temperatures on land to a widespread diagnostic array of measurements for all major Earth system components – whether deep ocean circulation or top of atmosphere radiation, sea ice extent or terrestrial carbon store, biodiversity of tropical forests or urban heat islands. Climate records have been painstakingly reconstructed for the past to set the present and future decades in context. Researchers now place more of their data in the public domain. With this 'digital Earth' we have assembled more detailed models and begun to understand planetary limits. This includes the 'safe operating space for humanity' (Steffen et al., 2015), whether it be GHG emissions or global freshwater use. In theory, better measurement offers a basis for better management.

Second, following the 2015 Conference of the Parties in Paris, there has literally been a paradigm shift in the approach to reducing global GHG emissions. The top-down framework adopted for the Kyoto Protocol has gone. This required international consensus about targets for decarbonising economies and yielded only modest results. Now the bottom-up framework of Intended Nationally Determined Contributions (INDCs) gives individual nations the responsibility to set their own priorities for mitigation and adaptation, recognising local development goals and circumstances. Accountability can be further devolved within national plans of action right down to communities, households and individuals. These offerings are judged in terms of their fairness, equity and ambition. Whether the collective commitment and pace of action will be sufficient to avoid dangerous climate change is, however, debatable. Early signs were not promising – by July 2016 only 19 (out of 197) Parties had ratified the Paris Agreement, amounting to just 0.18% of global GHG emissions.[5]

Finally, improved human security – put simply as freedom from want and freedom from fear – has emerged as a purposeful outcome of climate services. The necessary information, tools and principles are coming together through more widespread cross-disciplinary

---

[5] Although the Paris Agreement came into force on 4 November 2016, Donald Trump has threatened to withdraw the United States from the framework and to sweep away Obama's policies on emissions reduction.

working. Rather than being told, climate information users are more and more being listened to by climate information suppliers. Climate service products are increasingly being tailored to the local social and cultural context of decision-makers. Despite large uncertainty in the climate change outlook, adaptation options can still be tested, and resilience strengthened now. This all requires mainstreaming of *climate change in practice*. It is hoped that the questions and exercises provided by this book point to how this might be done.

18.2 GROUP EXERCISE: MINI COPENHAGEN CONSENSUS PANEL

NOTES

The purpose of this activity is to conduct a 'mini' Copenhagen Consensus Panel to consider investment options and potential opportunity costs associated with various climate-related activities. During the actual panel, experts start with a list of global challenges and are asked to identify costs and benefits of solutions. Challenges that have been covered so far include biodiversity loss, chronic disease, hunger and malnutrition, climate change, water and sanitation, infectious disease, population growth, trade barriers and corruption. Participants are given $75 billion to allocate to their list of priorities in order to ameliorate or eradicate the challenge.

In this variant, participants consider a more modest investment of £200 million – approximately the operating budget and cost of the UK Met Office Hadley Centre for five years including the expense of a new supercomputer (see Chapter 3). If helpful, further background information on the services provided by the Hadley Centre can be researched by the group beforehand.[6] Likewise, the suggested list of investment options can be customised to address priorities in other regions. For an added twist, group members could be assigned different roles (e.g. climate scientist, development officer, disaster relief agent, conservationist or consumer) and instructed to make the case for their own priority investment(s).

---

[6] Review of climate science advice to Government and Met Office Hadley Centre role, governance and resourcing. www.gov.uk/government/uploads/system/uploads/attachment_data/file/286122/10–1290-review-of-climate-science-advice.pdf

Mini Copenhagen Consensus Panel Worksheet

What criteria can be used to compare benefits arising from contrasting investments in social and environmental protection against climate variability and change? What indices might be used to measure these benefits? [Hint: review some of the investment proposals presented by Bjorn Lomborg.[7]]

Can climate change really be separated from all the investment proposals described by Lomborg?

Rank the following 'climate investments' in terms of priority, assumed benefits, and/or value for money. Each option costs approximately £200 million (2014 prices):

Options	Rank
Met Office Hadley Centre operating and supercomputer budget (5 years)	
Build 20 km of sea wall to protect settlements from coastal erosion or flooding	
Expand the National Forest by 20,000 hectares	
Build 4000 cyclone shelters in Bangladesh (protecting up to 4 million people and 2 million cattle)	
Award 4000 MSc scholarships to study global environmental change (annually for 5 years)	
Restore 200 km of inland waterways and canals	
Install free solar panels in 28,000 homes (a town the size of Loughborough)	
Construct sand dams to meet the water needs of 20 million people in the Sahel	
Build a 10 km mass-transit (tram) system in a UK city of your choice	
Install free loft insulation in 500,000 homes (saving 0.5 M tonnes CO_2 per year)	

What other far-reaching options could be added to the above list (to strengthen resilience to climate variability and change)?

---

[7] https://www.ted.com/talks/bjorn_lomborg_sets_global_priorities#t-322809

LEARNING OUTCOMES

- To be aware of the various pathways connecting climate to other global challenges of water, food, energy and regional security.
- To appreciate that multiple global challenges may coincide in time and space, resulting in a 'perfect storm'.
- To recognise that the nexus concept helps to identify policies and actions that can yield beneficial outcomes for more than one sector.

18.3 QUESTIONS TO TEST YOUR UNDERSTANDING

What is a climate nexus? Explain your answer with reference to the climate–energy–poverty nexus.

Via what pathways might climate shocks spark civil unrest and regional conflict?

What will be the most significant challenges faced by a world with 9 billion people in 2050?

18.4 FURTHER READING

Goodchild, M.F., Guo, H., Annoni, A., et al. 2012. Next-generation Digital Earth. *Proceedings of the National Academy of Sciences*, **109**, 11088–11094.

Lomborg, B. 2014. *How to Spend $75 Billion to Make the World a Better Place*, Second Edition. Copenhagen Consensus Center, Denmark.

Steffen, W., Richardson, K., Rockström, J., et al. 2015. Planetary boundaries: Guiding human development on a changing planet. *Science*, **347**, 736–746.

18.5 OTHER RESOURCES

Copenhagen Consensus Panel http://ranksmartsolutions.com/#gather-your-expert-panel [accessed 18/07/16]

IUCN Red List of Threatened Species www.iucnredlist.org/ [accessed 18/07/16]

Pew Research Center www.pewresearch.org/ [accessed 18/07/16]

UN FAO AQUASTAT www.fao.org/nr/water/aquastat/main/index.stm
    [accessed 18/07/16]
UN-Water www.unwater.org/home/en/ [accessed 18/07/16]
World Population Clock www.worldometers.info/world-population/
    [accessed 18/07/16]
WWF Living Planet Report http://wwf.panda.org/about_our_earth/
    all_publications/living_planet_report/ [accessed 18/07/16]

References

Abdallah, B., Mouhous-Voyneau, N. and Denoeux, T. 2014. Combining statistical and expert evidence using belief functions: application to centennial sea level estimation taking into account climate change. *International Journal of Approximate Reasoning*, **55**, 341–354.

Adams, P., Hewitson, B., Vaughan, C., et al. 2015. Call for an ethical framework for climate services. *WMO Bulletin*, **64**, 51–54.

Adger, W.N., Arnell, N.W. and Tompkins, E.L. 2005. Successful adaptation to climate change across scales. *Global Environmental Change*, **15**, 77–86.

Adger, W.N., Dessai, S., Goulden, M., et al. 2009. Are there social limits to adaptation to climate change? *Climatic Change*, **93**, 335–354.

Adger, W.N., Huq, S., Brown, K., Conway, D. and Hulme, M. 2003. Adaptation to climate change in the developing world. *Progress in Development Studies*, **3**, 179–195.

Adler, R.F., Huffman, G.J., Chang, A., et al. 2003. The Version 2 Global Precipitation Climatology Project (GPCP) Monthly Precipitation Analysis (1979–present). *Journal of Hydrometeorology*, **4**, 1147–1167.

Aklilu, Y. and Wekesa, M. 2002. *Drought, Livestock and Livelihood Lessons from the 1999–2001 Emergency Response in the Pastoral Sector in Kenya.* Humanitarian Practice Network Paper 40. Overseas Development Institute, London.

Alcamo, J., Flörke, M. and Märker, M. 2007. Future long-term changes in global water resources driven by socio-economic and climatic changes. *Hydrological Sciences Journal*, **52**, 247–275.

Allen, M.R. 2003. Liability for climate change. *Nature*, **421**, 892.

Allen, M.R., Frame, D.J., Huntingford, C., et al. 2009. Warming caused by cumulative carbon emissions towards the trillionth tonne. *Nature*, **458**, 1163–1166.

Amaral, L.P., Martins, N. and Gouveia, J.B. 2015. Quest for a sustainable university: a review. *International Journal of Sustainability in Higher Education*, **16**, 155–172.

Amengual, A., Homar, V., Romero, R., et al. 2014. Projections of heat waves with high impact on human health in Europe. *Global Planetary Change*, **119**, 71–84.

Amos, J. 2015. 'Next Pinatubo' a test of geoengineering. BBC News Science and Environment, 15 February 2015. www.bbc.co.uk/news/science-environment-31475761 [accessed 21/4/16].

Anderegg, W.R.L., Prall, J.W., Harold, J. and Schneider, S.H. 2010. Expert credibility in climate change. *Proceedings of the National Academy of Sciences of the United States of America*, **107**, 12107–12109.

Anderson, K. 2012. Climate change going beyond dangerous: brutal numbers and tenuous hope. *Development Dialogue*, **61**, 16–39.

Annamalai, H., Hamilton, K. and Sperber, K.R. 2007. The South Asian summer monsoon and its relationship with ENSO in the IPCC AR4 simulations. *Journal of Climate*, **20**, 1071–1092.

Antwi-Agyei, P., Dougill, A.J. and Stringer, L.C. 2014. Barriers to climate change adaptation: evidence from northeast Ghana in the context of a systematic literature review. *Climate and Development*, **14**, 1615–1626.

Appelquist, L.R. and Balstrøm, T. 2014. Application of the Coastal Hazard Wheel methodology for coastal multi-hazard assessment and management in the state of Djibouti. *Climate Risk Management*, **3**, 79–95.

Applegate, P.J. and Keller, K. 2015. How effective is albedo modification (solar radiation management geoengineering) in preventing sea-level rise from the Greenland Ice Sheet? *Environmental Research Letters*, **10**, 084018.

Arnell, N.W. and Gosling, S.N. 2016. The impacts of climate change on river flood risk at the global scale. *Climatic Change*, **134**, 387–401.

Arrhenius, S. 1908. *Worlds in the Making: The Evolution of the Universe*. Harper, New York.

Averchenkova, A. and Bassi, S. 2016. *Beyond the Targets: Assessing the Political Credibility of Pledges for the Paris Agreement*. London School of Economics, London.

Baettig, M., Wild, M. and Imboden, D.M. 2007. A climate change index: where climate change may be most prominent in the 21st century. *Geophysical Research Letters*, **34**, L01705.

Bakker, P., Masson-Delmotte, V., Martrat, B., et al. 2014. Temperature trends during the Present and Last Interglacial periods: a multi-model-data comparison. *Quaternary Science Reviews*, **99**, 224–243.

Bamber, J.L. and Aspinall, W.P. 2013. An expert judgement assessment of future sea level rise from ice sheets. *Nature Climate Change*, **3**, 434–427.

Barney, J.N. and DiTomasco, J.M. 2008. Nonnative species and bioenergy: are we cultivating the next invader? *BioScience*, **58**, 64–70.

Bastagli, F. and Harman, L. 2015. *The Role of Index-Based Triggers in Social Protection Response*. Overseas Development Institute, London.

Bauer, E., Claussen, M., Bovkin, V. and Huenerbein, A. 2003. Assessing climate forcings of the Earth system for the past millennium. *Geophysical Research Letters*, **30**, 1276.

Bealey, C. and Rayaleh, H.A. 2006. Understanding the status of the Djibouti Francolin. *Annual Review of the World Pheasant Association 2005/2006*, 20.

Beck, S. 2012. The challenges of building cosmopolitan climate expertise: the case of Germany. *WIREs Climate Change*, **3**, 1–17.

Beddington, J. 2009. *Food, Energy, Water and the Climate: A Perfect Storm of Global Events?* Government Office for Science, London.

Bellenger, H., Guilyardi, E., Leloup, J., Lengaigne, M. and Vialard, J. 2013. ENSO representation in climate models: from CMIP3 to CMIP5. *Climate Dynamics*, **42**, 1999–2018.

Benjaminsen, T.A., Alinon, K., Buhaug, H. and Buseth, J.T. 2012. Does climate change drive land-use conflicts in the Sahel? *Journal of Peace Research*, **49**, 97–111.

Berger, W.H. 2013. Milankovitch tuning of deep-sea records: implications for maximum rates of change of sea level. *Global and Planetary Change*, **101**, 131–143.

Berrittella, M., Certa, A., Enea, M. and Zito, P. 2008. Transport policy and climate change: how to decide when experts disagree. *Environmental Science and Policy*, **11**, 307–314.

Blokker, E.J.M., van Osch, A.M., Hogveen, R. and Mudde, C. 2013. Thermal energy from drinking water and cost benefit analysis for an entire city. *Journal of Water and Climate Change*, **4**, 11–16.

Bloodhart, B., Maibach, E., Myers, T. and Zhao, X.Q. 2015. Local climate experts: the influence of local TV weather information on climate change perceptions. *PLOS ONE*, **10**, e0141526.

Bodenstein, L. 2010. Regarding Anderegg et al. and climate change credibility. *Proceedings of the National Academy of Sciences of the United States of America*, **107**, E188.

Boucher, O., Lowe, J.A. and Jones, C.D. 2009. Implications of delayed actions in addressing carbon dioxide emission reduction in the context of geo-engineering. *Climatic Change*, **92**, 261–273.

Bowman, T.E., Maibach, E., Mann, M.E., et al. 2010. Time to take action on climate communication. *Science*, **330**, 1044.

Boykoff, M.T. 2007. From convergence to contention: United States mass media representations of anthropogenic climate change science. *Transactions of the Institute of British Geographers*, **32**, 477–489.

Boykoff, M.T. 2008a. The cultural politics of climate change discourse in UK tabloids. *Political Geography*, **27**, 549–569.

Boykoff, M.T. 2008b. Lost in translation? United States television news coverage of anthropogenic climate change, 1995–2004. *Climatic Change*, **86**, 1–11.

Boykoff, M.T. 2013. Public enemy no. 1?: understanding media representations of outlier views on climate change. *American Behavioural Scientist*, **57**, 796–817.

Boykoff, M.T. and Boykoff, J.M. 2004. Balance as bias: global warming and the US prestige press. *Global Environmental Change*, **14**, 125–136.

Boykoff, M., Daly, M., Gifford, L., et al. 2016. Australian newspaper coverage of climate change or global warming, 2004–2016. Center for Science and Technology Policy Research, Cooperative Institute for Research in Environmental Sciences, University of Colorado [accessed 1/6/16].

Brecht, H., Dasgupta, S., Laplante, B., Murray, S. and Wheeler, D. 2012. Sea-level rise and storm surges: high stakes for a small number of developing countries. *Journal of Environment and Development*, **21**, 120–138.

Brekke, L.D., Maurer, E.P., Anderson, J.D., et al. 2009. Assessing reservoir operations risk under climate change. *Water Resources Research*, **45**, W04411.

Brooks, N., Adger, W.N. and Kelly, P.M. 2005. The determinants of vulnerability and adaptive capacity at the national level and the implications for adaptation. *Global Environmental Change*, **15**, 151–163.

Brown, C. and Wilby, R.L. 2012. An alternate approach to assessing climate risks. *Eos*, **92**, 401–403.

Brown, C., Werick, W., Leger, W. and Fay, D. 2011. A decision-analytic approach to managing climate risks: application to the Upper Great Lakes. *Journal of the American Water Resources Association*, **47**, 524–534.

Bryden, H.L., Longworth, H.R. and Cunningham, S.A. 2005. Slowing of the Atlantic meridional overturning circulation at 25° N. *Nature*, **438**, 655–657.

Buhaug, H. 2010. Climate not to blame for African civil wars. *Proceedings of the National Academy of Sciences of the United States of America*, **107**, 16477–16482.

Buhaug, H., Gleditsch, N.P. and Theisen, O.M. 2008. Implications of climate change for armed conflict. *Social Dimensions of Climate Change*, World Bank, Washington DC.

Burke, M.B., Miguel, E., Satyanath, S., Dykema, J.A. and Lobell, D.B. 2009. Warming increases the risk of civil war in Africa. *Proceedings of the National Academy of Sciences of the United States of America*, **106**, 20670–20674.

Burkett, V.R., Suarez, A.G., Bindi, M., et al. 2014. Point of departure. In *Climate Change 2014: Impacts, Adaptation, and Vulnerability. Part A: Global and Sectoral Aspects. Contribution of Working Group II to the Fifth Assessment Report of the Intergovernmental Panel on Climate Change* [Field, C.B., Barros, V.R., Dokken, D.J., et al. (eds.)]. Cambridge University Press, Cambridge.

Burrows, K. and Kinney, P.L. 2016. Exploring the climate change, migration and conflict nexus. *International Journal of Environmental Research and Public Health*, **13**, 443.

Bybee, R., McCrae, B. and Laurie, R. 2009. PISA 2006: an assessment of scientific literacy. *Journal of Research in Science Teaching*, **46**, 865–883.

Byers, E.A., Qadrdan, M., Kilsby, C.G., et al. 2015. Cooling water for Britain's future electricity supply. *Proceedings of the Institution of Civil Engineers: Energy*, **168**, 188–204.

Cahill, N., Rahmstorf, S. and Parnell, A.C. 2015. Change points of global temperature. *Environmental Research Letters*, **10**, 084002.

Callendar, G.S. 1938. The artificial production of carbon dioxide and its influence on temperature. *Quarterly Journal of the Royal Meteorological Society*, **64**, 223–240.

Camargo, S.J. and Sobel, A.H. 2005. Western North Pacific tropical cyclone intensity and ENSO. *Journal of Climate*, **18**, 2996–3006.

Capstick, S.B. and Pidgeon, N.F. 2014. What is climate change scepticism? Examination of the concept using a mixed methods study of the UK public. *Global Environmental Change*, **24**, 389–401.

Carbon Trust, 2006. *The Carbon Emissions Generated in All That We Consume*. The Carbon Trust, London.

Carbon Trust, 2014. *Homeworking: Helping Businesses Cut Costs and Reduce Their Carbon Footprint*. The Carbon Trust, London.

Carlton, J.S. and Jacobson, S.K. 2016. Using expert and non-expert models of climate change to enhance communication. *Environmental Communication*, **10**, 1–24.

Carlton, J.S., Mase, A.S., Knutson, C.L., et al. 2016. The effects of extreme drought on climate change beliefs, risk perceptions, and adaptation attitudes. *Climatic Change*, **135**, 211–226.

Carvalho, A. and Burgess, J. 2005. Cultural circuits of climate change in UK broadsheet newspapers, 1985–2003. *Risk Analysis*, **25**, 1457–1469.

Casanueva, A., Frías, M.D., Herrera, S., et al. 2014. Statistical downscaling of climate impact indices: testing the direct approach. *Climatic Change*, **127**, 547–560.

Castro, C.L., Pielke, R.A. and Leoncini, G. 2005. Dynamical downscaling: assessment of value retained and added using the regional atmospheric modelling system (RAMS). *Journal of Geophysical Research: Atmospheres*, **110**, D05108.

Chapagain, A.K. and Hoekstra, A.Y. 2008. The global component of freshwater demand and supply: an assessment of virtual water flows between nations as a result of trade in agricultural and industrial products. *Water International*, **33**, 19–32.

Chaudhary, A. and Kastner, T. 2016. Land use biodiversity impacts embodied in international food trade. *Global Environmental Change*, **38**, 195–204.

Chen, X. and Tung, K.-K. 2014. Varying planetary heat sink to global-warming slowdown and acceleration. *Science*, **345**, 897–903.

Chevallier, P., Pouyard, M., Bolgov, M., et al. 2012. Trends for snow cover and river flows in the Pamirs (Central Asia). *Hydrology and Earth System Sciences Discussions*, **9**, 29–64.

Christen, A., Coops, N.C., Crawford, C.R., et al. 2011. Validation of modeled carbon-dioxide emissions from an urban neighbourhood with direct eddy-covariance measurements. *Atmospheric Environment*, **45**, 6057–6069.

Cicerone, R.J. 2006. Geoengineering: encouraging research and overseeing implementation. *Climatic Change*, **77**, 221–226.

Collins, H.M. and Evans, R. 2002. The third wave of science studies: studies of expertise and experience. *Social Studies of Science*, **32**, 235–296.

Collins, W.J., Bellouin, N., Doutriaux-Boucher, M., et al. 2011. Development and evaluation of an Earth-system model: HadGEM2. *Geoscientific Model Development*, **4**, 1051–1075.

Comité National pour l'Environnement (CNE), 1991. *Rapport National Environnement*. Secrétariat Technique du Comité National pour l'Environnement, Djibouti.

Committee on Climate Change (CCC), 2015a. *The Fifth Carbon Budget: The Next Step Towards a Low-Carbon Economy*. Committee on Climate Change, London.

Committee on Climate Change (CCC), 2015b. *Sectoral Scenarios for the Fifth Carbon Budget*. Technical Report. Committee on Climate Change, London.

Committee on Climate Change (CCC), 2015c. *Progress in Preparing for Climate Change*. Adaptation Sub Committee, London.

Committee on Climate Change (CCC), 2016. *UK Climate Change Risk Assessment 2017. Synthesis Report: Priorities for the Next Five Years*. Committee on Climate Change, London.

Contamin, R. and Ellison, A.M. 2009. Indicators of regime shifts in ecological systems: what we need to know and when do we need to know it? *Ecological Applications*, **19**, 799–816.

Cook, E.R., Seager, R., Heim, R.R., et al. 2009. Megadroughts in North America: placing IPCC projections of hydroclimatic change in a long-term palaeoclimate context. *Journal of Quaternary Science*, **25**, 48–61.

Cook, J., Nuccitelli, D., Green, S.A., et al. 2013. Quantifying the consensus on anthropogenic global warming in the scientific literature. *Environmental Research Letters*, **8**, 024024.

Corner, A. and Pidgeon, N. 2010. Geoengineering the climate: the social and ethical implications. *Environment: Science and Policy for Sustainable Development*, **52**, 24–37.

Corner, A. and Pidgeon, N. 2015. Like artificial trees? The effect of framing by natural analogy on public perceptions of geoengineering. *Climatic Change*, **130**, 425–438.

Covey, C., AchutaRao, K.M., Cubasch, U., et al. 2003. An overview of results from the Coupled Model Intercomparison Project. *Global and Planetary Change*, **37**, 103–133.

Crawford, E. 1997. Arrhenius' 1896 model of the greenhouse effect in context. *Ambio*, **26**, 6-11.

Crawford, T., Betts, N.L. and Favis-Mortlock, D.T. 2007. GCM grid box choice and predictor selection associated with statistical downscaling of daily precipitation over Northern Ireland. *Climate Research*, **34**, 145–160.

Cressey, D. 2012. Cancelled project spurs debate over geoengineering patents. *Nature*, **485**, 429.

Crook, J.A. and Forster, P.M. 2011. A balance between radiative forcing and climate feedback in the modelled 20th century temperature response. *Journal of Geophysical Research: Atmospheres*, **116**, D17108.

Crook, J.A., Jackson, L.S., Osprey, S.M. and Forster, P.M. 2015. A comparison of temperature and precipitation responses to different Earth radiation management geoengineering schemes. *Journal of Geophysical Research: Atmospheres*, **120**, 9352–9373.

Crowley, T.J. 2000. Causes of climate change over the past 1000 years. *Science*, **289**, 270–277.

Crutzen, P.J. 2006. Albedo enhancement by stratospheric sulphate injections: a contribution to resolve a policy dilemma. *Climatic Change*, **77**, 211–219.

Cubasch, U., Wuebbles, D., Chen, D., et al. 2013. Introduction. In *Climate Change 2013: The Physical Science Basis. Contribution of Working Group I to the Fifth Assessment Report of the Intergovernmental Panel on Climate Change* [Stocker, T.F., Qin, D., Plattner, G.-K., et al. (eds.)]. Cambridge University Press, Cambridge.

Curry, J.A. 2014a. Climate science: uncertain temperature trends. *Nature Geoscience*, **7**, 83–84.

Curry, J.A. 2014b. End of climate exceptionalism. https://judithcurry.com/ 2014/04/04/end-of-climate-exceptionalism/ [accessed 16/06/16]

Curry, J.A. 2015. Has NOAA 'busted' the pause in global warming? https:// judithcurry.com/2015/06/04/has-noaa-busted-the-pause-in-global-warming/ [accessed 29/06/16]

Curry, J.A. and Webster, P.J. 2011. Climate science and the uncertainty monster. *Bulletin of the American Meteorological Society*, **92**, 1667–1682.

Dahan-Dalmedico, A. 2008. Climate expertise: between scientific credibility and geopolitical imperatives. *Interdisciplinary Science Reviews*, **33**, 71–81.

Dai, A. 2011. Drought under global warming: a review. *Wiley Interdisciplinary Science Reviews*, **2**, 45–65.

Dai, A. 2013. Increasing drought under global warming in observations and models. *Nature Climate Change*, **3**, 52–58.

d'Amour, C.B., Wenz, L., Kalkul, M., Steckel, J.C. and Creutzig, F. 2016. Teleconnected food supply shocks. *Environmental Research Letters*, **11**, 035007.

Dasgupta, S., Laplante, B., Murray, S. and Wheeler, D. 2009. *Climate Change and the Future Impacts of Storm-Surge Disasters in Developing Countries*. Working Paper 182, Center for Global Development, Washington DC.

Davey, C.A. and Pielke, R.A. 2005. Microclimate exposures of surface-based weather stations: implications for the assessment of long-term temperature trends. *Bulletin of the American Meteorological Society*, **86**, 497–504.

Davin, E.L., Seneviratne, S.I., Cias, P., Olioso, A. and Wange, T. 2014. Preferential cooling of hot extremes from cropland albedo management. *Proceedings of the National Academy of Sciences of the United States of America*, **111**, 9757–9761.

Davis, M. 2001. *Late Victorian Holocausts: El Niño Famines and the Making of the Third World*. Verso, London.

Davis, S.J. and Caldeira, K. 2010. Consumption-based accounting of CO_2 emissions. *Proceedings of the National Academy of Sciences of the United States of America*, **107**, 5687–5692.

Dawson, R.J. 2011. Potential pitfalls on the transition to more sustainable cities and how they might be avoided. *Carbon Management*, **2**, 175–188.

De Martino, B., Kumaran, D., Seymour, B. and Dolan, R.J. 2006. Frames, biases, and rational decision-making in the human brain. *Science*, **313**, 684–687.

de Wit, M. and Stankiewicz, J. 2006. Changes in surface water supply across Africa with predicted climate change. *Science*, **311**, 1917–1921.

De'ath, G., Lough, J.M. and Fabricius, K.E. 2009. Declining coral calcification on the Great Barrier Reef. *Science*, **323**, 116–119.

Defra, 2012. *Summary of the Key Findings from the UK Climate Change Risk Assessment 2012*. HM Government, London.

Delta Commission, 2008. *Working Together with Water: A Living Land Builds for its Future*. Delta Commission, The Netherlands.

DeNicola, E., Aburizaiza, O.S., Siddique, A., Khwaja, H. and Carpenter, D.O. 2015. Climate change and water scarcity: the case of Saudi Arabia. *Annals of Global Health*, **81**, 342–353.

Deser, C., Phillips, A.S., Alexander, M.A. and Smoliak, B. 2014. Projecting North American climate over the next 50 years: uncertainty due to internal variability. *Journal of Climate*, **27**, 2271–2296.

Dessai, S., Adger, W.N., Hulme, M., et al. 2004. Defining and experiencing dangerous climate change. *Climatic Change*, **64**, 11–25.

Dessai, S., Lu, X. and Risbey, J.S. 2005. On the role of climate scenarios for adaptation planning. *Global Environmental Change*, **15**, 87–97.

Di Luca, A., de Elia, R. and Laprise, R. 2013. Potential for small scale added value of RCM's downscaled climate change signal. *Climate Dynamics*, **40**, 601–618.

Diaz-Nieto, J. and Wilby, R.L. 2005. A comparison of statistical downscaling and climate change factor methods: impacts on low flows in the River Thames, United Kingdom. *Climatic Change*, **69**, 245–268.

Diffenbaugh, N.S. and Scherer, M. 2013. Likelihood of July 2012 U.S. temperatures in preindustrial and current forcing regimes. *Bulletin of the American Meteorological Society*, **94**, S6–S9.

Direction de L'Aménagement du Territoire et de l'Environnement (DATE), 2001. *Communication Nationale Initiale de la Republique de Djibouti a la Convention Cadres des Nations Unies sur les Changements Climatiques*. Ministère de l'Habitat, de l'Urbanisme, de l'Environnement et de l'Aménagement du Territoire, République de Djibouti.

Direction de L'Aménagement du Territoire et de l'Environnement (DATE), 2006. *Programme d'Action National d'Aptation aux Changements Climatiques*. Ministère de l'Habitat, de l'Urbanisme, de l'Environnement et de l'Aménagement du Territoire, République de Djibouti.

Dobler, C., Hagemann, S., Wilby, R.L. and Stötter, J. 2012. Quantifying different sources of uncertainty in hydrological projections at the catchment scale. *Hydrology and Earth Systems Science*, **16**, 4343–4360.

Dole, R., Hoerling, M., Perlwitz, J., et al. 2011. Was there a basis for anticipating the 2010 Russian heatwave? *Geophysical Research Letters*, **38**, L06702.

Doll, R., Peto, R., Boreham, J. and Sutherland, I. 2004. Mortality in relation to smoking: 50 years' observations on male British doctors. *British Medical Journal*, **328**, 1519–1527.

Doria, M.D., Boyd, E., Tompkins, E.L. and Adger, W.N. 2009. Using expert elicitation to define successful adaptation to climate change. *Environmental Science and Policy*, **12**, 810–819.

Doulton, H. and Brown, K. 2009. Ten years to prevent catastrophe? Discourses of climate change and international development in the UK press. *Global Environmental Change*, **19**, 191–202.

Druckman, A. and Jackson, T. 2009. The carbon footprint of UK households 1990–2004: a socio-economically disaggregated, quasi-multi-regional input-output model. *Ecological Economics*, **68**, 2066–2077.

Dutra, E., Magnusson, L., Wetterhall, F., et al. 2013. The 2010–2011 drought in the Horn of Africa in ECMWF reanalysis and seasonal forecast products. *International Journal of Climatology*, **33**, 1720–1729.

Easterling, D.R. and Wehner, M.F. 2009. Is the climate warming or cooling? *Geophysical Research Letters*, **36**, L08706.

Economist, 2014. Climate change: in the balance. *The Economist*, 5 April 2014.

Economist, 2016. Particular about particulates. *The Economist*, 16 January 2016, 56–57.

Edenhofer, O., Pichs-Madruga, R., Sokona, Y., et al. 2014. Technical Summary. In *Climate Change 2014: Mitigation of Climate Change. Contribution of Working Group III to the Fifth Assessment Report of the Intergovernmental Panel on Climate Change* [Edenhofer, O., Pichs-Madruga, R., Sokona, Y., et al. (eds.)]. Cambridge University Press, Cambridge.

Edwards, P.N. 2011. History of climate modelling. *WIREs Climate Change*, **2**, 128–139.

Eggers, J., Tröltzsch, K., Falcucci, A., et al. 2009. Is biofuel harming biodiversity in Europe? *Global Change Biology: Bioenergy*, **1**, 18–34.

England, M.H., McGregor, S., Spence, P., et al. 2014. Recent intensification of wind-driven circulation in the Pacific and the ongoing warming hiatus. *Nature Climate Change*, **4**, 222–227.

Engler, S., Mauelshagen, F., Werner, J. and Luterbacher, J. 2013. The Irish famine of 1740–1741: famine vulnerability and "climate migration". *Climate of the Past*, **9**, 1161–1179.

Entman, R.M. 1993. Framing: toward clarification of a fractured paradigm. *Journal of Communication*, **43**, 51–58.

Environment Agency, 2011. *Advice for Flood and Coastal Erosion Risk Management*. Environment Agency, Bristol.

Environment Agency, 2012. *Getting Ready for Climate Change by Creating Riparian Shade*. Environment Agency, Bristol.

Ereaut, G. and Segnit, N. 2006. *Warm Words: How Are We Telling the Climate Story and Can We Tell It Better?* Institute for Public Policy Research, London.

Esmaeili, H. 2010. International law responses to climate change. In Lever-Tracy, C. (ed.), *Routledge Handbook of Climate Change and Society*. Routledge, Abingdon.

Estrada, F., Guerrero, V.M., Gay-García, C. and Martínez-López, B. 2013. A cautionary note on automated downscaling methods for climate change. *Climate Change*, **120**, 263–276.

European Bank for Reconstruction and Development (EBRD), 2011. *Tajikistan Pilot Programme for Climate Resilience (PPCR) Project A4 - Improving the Climate Resilience of Tajikistan's Hydropower Sector (Final report)*. Acclimatise and Sinclair Knight Mertz, Oxford.

European Bank for Reconstruction and Development (EBRD), 2015. *Qairokkum Hydropower: Planning Ahead for a Changing Climate*. Sustainable Energy Initiative, EBRD, London.

Farbotko, C. and Lazrus, H. 2012. The first climate refugees? Contesting global narratives of climate change in Tuvalu. *Global Environmental Change*, **22**, 382–390.

Fargione, J., Hill, J., Tilman, D., Polasky, S. and Hawthorne, P. 2008. Land clearing and the biofuel carbon debt. *Science*, **319**, 1235–1238.

Feng, X., Porporato, A. and Rodriguez-Iturbe, I. 2013. Changes in rainfall seasonality in the tropics. *Nature Climate Change*, **3**, 811–815.

Ferkany, M. 2015. Is it arrogant to deny climate change or is it arrogant to say it is arrogant? Understanding arrogance and cultivating humility in climate change discourse and education. *Environmental Values*, **24**, 705–724.

Field, C.B., Barros, V.R., Mach, K.J., et al. 2014. Technical summary. In *Climate Change 2014: Impacts, Adaptation, and Vulnerability. Part A: Global and Sectoral Aspects. Contribution of Working Group II to the Fifth Assessment Report of the Intergovernmental Panel on Climate Change* [Field, C.B., Barros, V.R., Dokken, D.J. et al. (eds.)]. Cambridge University Press, Cambridge.

Fields, D., Kochnakyan, A., Stuggins, G. and Besant-Jones, J. 2012. *Tajikistan's Winter Energy Crisis: Electricity Supply and Demand Alternatives*. World Bank, Washington DC.

Fischer, E.M. and Schär, C. 2010. Consistent geographical patterns of changes in high-impact European heatwaves. *Nature Geoscience*, **3**, 398–403.

Fisher, Z.S., Cartwright, S., Bealey, C., et al. 2009. The Djibouti francolin and juniper forest in Djibouti: the need for both ecosystem and species-specific conservation. *Oryx*, **43**, 542–551.

Flato, G.J., Marotzke, B., Abiodun, P., et al. 2013. Evaluation of climate models. In *Climate Change 2013: The Physical Science Basis. Contribution of Working Group I to the Fifth Assessment Report of the Intergovernmental Panel on Climate Change* [Stocker, T.F., Qin, D., Plattner, G.-K., et al. (eds.)]. Cambridge University Press, Cambridge.

Foresight Futures, 2011. *The Future of Food and Farming: Challenges and Choices for Global Sustainability*. Government Office for Science, London.

Fowler, H.J. and Wilby, R.L. 2007. Editorial: Beyond the downscaling comparison study. *International Journal of Climatology*, **27**, 1543–1545.

Francis, P. 2004. The impact of UK households on the environment. In *Economic Trends 611*. Office of National Statistics, London.

Frigg, R., Smith, L.A., Stainforth, D.A. 2013. The myopia of imperfect climate models: the case of UKCP09. *Philosophy of Science*, **80**, 886–897.

Frijns, J. 2012. Towards a common carbon footprint assessment methodology for the water sector. *Water and Environment Journal*, **25**, 63–69.

Fung, F., Lopez, A. and New, M. 2011. Water availability in +2°C and +4°C worlds. *Transactions of the Royal Society A*, **369**, 99–116.

Funk, C., Hoell, A., Shukla, S., et al. 2014. Predicting East African spring droughts using Pacific and Indian Ocean sea surface temperature indices. *Hydrology and Earth System Sciences*, **18**, 4965–4978.

Funke, G. 2010. The Munich Re view on climate change litigation. In *Munich Re: Liability for climate change?* Munich Re, Munich, Germany.

Fuss, S., Canadell, J. G., Peters, G. P., et al. 2014. Betting on negative emissions. *Nature Climate Change*, **4**, 850–853.

Fuss, S., Reuter, W.H., Szolgayova, J. and Obersteiner, M. 2013. Optimal mitigation strategies with negative emission technologies and carbon sinks under uncertainty. *Climatic Change*, **118**, 73–87.

Füssel, H.-M. and Klein, R.J. 2006. Climate change vulnerability assessments: an evolution of conceptual thinking. *Climatic Change*, **75**, 301–329.

Gabriel, C.J. and Robock, A. 2015. Stratospheric geoengineering impacts on El Nino/Southern Oscillation. *Atmospheric Chemistry and Physics*, **15**, 11949–11966.

Gemenne, F., Barnett, J., Adger, W.N. and Dabelko, G.D. 2014. Climate and security: evidence, emerging risks, and a new agenda. *Climatic Change*, **123**, 1–9.

Gençsü, I. and Hino, M. 2015. *Raising Ambition to Reduce International Aviation and Maritime Emissions*. New Climate Economy, London and Washington DC.

Ghosh, S., Sharma, A., Arora, S. and Desouza, G. 2016. A geoengineering approach toward tackling tropical cyclones over the Bay of Bengal. *Atmospheric Science Letters*, **17**, 208–215.

Gleckler, P.J., Wigley, T.M.L., Santer, B.D., et al. 2006. Krakatoa's signature persists in the ocean. *Nature*, **439**, 675.

Gleick, P.H. 2014. Water, drought, climate change, and conflict in Syria. *Weather Climate and Society*, **6**, 331–340.

Gleick, P.H., Yolles, P. and Hatami, H. 1994. Water, war and peace in the Middle East. *Environment*, **36**, 6–42.

Goklany, I.M. 2015. *Carbon Dioxide: The Good News*. Global Warming Policy Foundation, London.

Gonzalez, A., Cardinale, B.J., Allington, G.R.H., et al. 2016. Estimating local biodiversity change: a critique of papers claiming no net loss of local diversity. *Ecology*, **97**, 1949–1960.

Grundmann, R. 2012. The legacy of Climategate: revitalizing or undermining climate science and policy? *WIREs Climate Change*, **3**, 281–288.

Grundmann, R. 2013. "Climategate" and the scientific ethos. *Science Technology Human Values*, **38**, 67–93.

Gustafson, W.I., Qian, Y. and Fast, J.D. 2011. Downscaling aerosols and the impact of neglected subgrid processes on direct aerosol radiative forcing for a representative global climate model grid spacing. *Journal of Geophysical Research: Atmospheres*, **116**, D13303.

Gynther, I., Waller, N. and Leung, L.K.-P. 2016. *Confirmation of the Extinction of the Bramble Cay Melomys Melomys Rubicola on Bramble Cay, Torres Strait: Results and Conclusions from a Comprehensive Survey in August–September 2014*. Unpublished report to the Department of Environment and Heritage Protection, Queensland Government, Brisbane.

Haensler, A., Hagemann, S. and Jacob, D. 2011. Dynamical downscaling of ERA40 reanalysis data over southern Africa: added value in the simulation of the seasonal rainfall characteristics. *International Journal of Climatology*, **31**, 2338–2349.

Hahn, M.B., Riederer, A.M and Foster, S.O. 2009. The Livelihood Vulnerability Index: a pragmatic approach to assessing risks from climate variability

and change – A case study in Mozambique. *Global Environmental Change*, **19**, 74–88.

Hajat, S., Kovats, R.S., Atkinson, R.W. and Haines, A. 2002. Impact of hot temperatures on death in London: a time series approach. *Journal of Epidemiology and Community Health*, **56**, 367–372.

Hallegatte, S. 2009. Strategies to adapt to an uncertain climate. *Global Environmental Change*, **19**, 240–247.

Hallegatte, S., Green, C., Nicholls, R.J. and Corfee-Morlot, J. 2013. Future flood losses in major coastal cities. *Nature Climate Change*, **3**, 802–806.

Hanjra, M.A. and Qureshi, M.E. 2010. Global water crisis and future food security in an era of climate change. *Food Policy*, **35**, 365–377.

Hardisty, J., Taylor, D.M. and Metcalfe, S.E. 1993. *Computerised Environmental Modelling: A Practical Introduction Using Excel*. Wiley, Chichester.

Harvey, C.A., Rakotobe, Z.L., Rao, N.S., et al. 2014. Extreme vulnerability of smallholder farmers in Madagascar. *Philosophical Transactions of the Royal Society B*, **369**, 20130089.

Haszeldine, R.S. 2009. Carbon capture and storage: how green can black be? *Science*, **325**, 1647–1652.

Haug, G.H., Gunther, D., Peterson, L.C., et al. 2003. Climate and the collapse of Maya civilization. *Science*, **299**, 1731–1735.

Hawkins, E. 2015a. Graphics: scrap rainbow colour scales. *Nature*, **519**, 291.

Hawkins, E. 2015b. Was there ever a 'pause'? www.climate-lab-book.ac.uk/2015/was-there-ever-a-pause/ [accessed 25/06/16]

Hawkins, E. and Sutton, R. 2010. The potential to narrow uncertainty in projections of regional precipitation change. *Climate Dynamics*, **37**, 407–418.

Hawkins, E., Smith, R.S., Gregory, J.M. and Stainforth, D.A. 2016. Irreducible uncertainty in near-term climate projections. *Climate Dynamics*, **46**, 3807–3819.

Haylock, M.R., Cawley, G.C., Harpham, C., Wilby, R.L. and Goodess, C.M. 2006. Downscaling heavy precipitation over the UK: a comparison of dynamical and statistical methods and their future scenarios. *International Journal of Climatology*, **26**, 1397–1415.

Heimann, M. and Reichstein, M. 2008. Terrestrial ecosystem carbon dynamics and climate feedbacks. *Nature*, **451**, 289–292.

Hellmuth, M.E., Mason, S.J., Vaughan, C., van Aalst, M.K. and Choularton, R. (eds.) 2011. *A Better Climate for Disaster Risk Management*. International Research Institute for Climate and Society (IRI), Columbia University, New York.

Heltberg, R., Siegel, P.B. and Jorgensen, S.L. 2009. Addressing human vulnerability to climate change: toward a 'no regret' approach. *Global Environmental Change*, **19**, 89–99.

Henson, R. 2014. *The Thinking Person's Guide to Climate Change*. The American Meteorological Society, Boston, MA.

Hochrainer, S. and Mechler, R. 2011. Natural disaster risk in Asian megacities: a case for risk pooling? *Cities*, **28**, 53–61.

Holderness, T., Barr, S., Dawson, R. and Hall, J. 2013. An evaluation of thermal Earth observation for characterizing urban heatwave event dynamics using the urban heat island intensity metric. *International Journal of Remote Sensing*, **34**, 864–884.

Honda, Y., Kondo, M., McGregor, G., et al. 2014. Heat-related mortality risk model for climate change impact projection. *Environmental Health and Preventive Medicine*, **19**, 56–63.

Hoornweg, D., Sugar, L. and Gómez, C.L.T. 2011. Cities and greenhouse gas emissions: moving forward. *Environment and Urbanization*, **23**, 207–227.

House of Commons Science and Technology Committee, 2014. *Communicating Climate Science. Eighth Report of Session 2013-14*, The Stationery Office, London.

House of Commons, 2015. *Future of Carbon Capture and Storage in the UK.* Energy and Climate Change Committee, The Stationery Office, London.

Houssein, I. and Jalludin, M. 1995. The salinity of Djibouti's aquifer. *Journal of African Earth Sciences*, **22**, 409–414.

Houssein, M.A. 2005. *Vulnérabilité des Zones Côtières aux Changements Climatiques: De l'État des Lieux à l'Élargissement de la Problématique.* Ministère de l'Habitat, de l'Urbanisme, de l'Environnement et de l'Aménagement du Territoire, République de Djibouti.

Howe, P.D., Thaker, J. and Leiserowitz, A. 2014. Public perceptions of rainfall change in India. *Climatic Change*, **127**, 211–225.

Hughes, T.P., Day, J.C. and Brodie, J. 2015. Securing the future of the Great Barrier Reef. *Nature Climate Change*, **5**, 508–511.

Hulme, M. 2007. Newspaper scare headlines can be counter-productive. *Nature*, **445**, 818.

Hulme, M. 2008. The conquering of climate: discourses of fear and their dissolution. *Geographical Journal*, **174**, 5–16.

Hulme, M. 2009. The communication of risk. In *Why We Disagree about Climate Change.* Cambridge University Press, Cambridge.

Hulme, M. and Dessai, S. 2008. Negotiating future climates for public policy: a critical assessment of the development of climate scenarios for the UK. *Environmental Science and Policy*, 11, 54–70.

Hunt, A.S.P., Wilby, R.L., Dale, N., Sura, K. and Watkiss, P. 2014. Embodied water imports to the UK under climate change. *Climate Research*, **59**, 89–101.

Idso, C.D., Carter, R.M. and Singer, S.F. 2015. *Why Scientists Disagree about Global Warming.* The Heartland Institute, Arlington Heights, IL.

Intergovernmental Panel on Climate Change (IPCC), 2001. *Climate Change 2001: Impacts, Adaptation and Vulnerability. Contribution of Working Group II to the Third Assessment Report of the Intergovernmental Panel on Climate Change* [McCarthy, J.J., Canziani, O.F., Leary, N.A., Dokken, D.J. and White, K.S. (eds.)]. Cambridge University Press, Cambridge.

Intergovernmental Panel on Climate Change (IPCC), 2006. *National Greenhouse Gas Inventories: Land Use, Land Use Change and Forestry.* Institute of Global Environmental Strategies, Hayama, Japan.

Intergovernmental Panel on Climate Change (IPCC), 2012. *Managing the Risks of Extreme Events and Disasters to Advance Climate Change Adaptation. A Special Report of Working Groups I and II of the Intergovernmental Panel on Climate Change* [Field, C.B., Barros, V., Stocker, T.F. et al. (eds.)]. Cambridge University Press, Cambridge.

Intergovernmental Panel on Climate Change (IPCC), 2013a. Summary for Policymakers. In *Climate Change 2013: The Physical Science Basis. Contribution of Working Group I to the Fifth Assessment Report of the Intergovernmental Panel on Climate Change* [Stocker, T.F., Qin, D., Plattner, G.-K. et al. (eds.)]. Cambridge University Press, Cambridge.

Intergovernmental Panel on Climate Change (IPCC), 2013b. Technical Summary. In *Climate Change 2013: The Physical Science Basis. Contribution of Working Group I to the Fifth Assessment Report of the Intergovernmental Panel on Climate Change* [Stocker, T.F., Qin, D. Plattner, G.-K. et al. (eds.)]. Cambridge University Press, Cambridge.

Intergovernmental Panel on Climate Change (IPCC), 2014. Summary for Policymakers. In *Climate Change 2014: Mitigation of Climate Change. Contribution of Working Group III to the Fifth Assessment Report of the Intergovernmental Panel on Climate Change* [Edenhofer, O., Pichs-Madruga, R., Sokona, Y. et al. (eds.)]. Cambridge University Press, Cambridge.

International Organization for Standardization (ISO), 2013. *Greenhouse Gases – Carbon Footprints of Products – Requirements and Guidelines for Quantification and Communication.* ISO/TS 14067:2013. ISO, Geneva, Switzerland.

Ioannidis, J.P.A. 2013. Informed consent, big data, and the oxymoron of research that is not research. *The American Journal of Bioethics*, **13**, 40–42.

Irvine, P.J., Ridgwell, A. and Lunt, D.J. 2011. Climatic effects of surface albedo geoengineering. *Journal of Geophysical Research: Atmospheres*, **116**, D24112.

Jacob, D., Peterson, J., Eggert, B., et al. 2014. EURO-CORDEX: new high-resolution climate change projections for European impact research. *Regional Environmental Change*, **2**, 563–578.

Jacques, P.J. 2012. A general theory of climate denial. *Global Environmental Politics*, **12**, 9-17.

Jacques, P.J., Dunlap, R. and Freeman, M. 2008. The organisation of denial: Conservative think tanks and environmental scepticism. *Environmental Politics*, **17**, 349–385.

Janko, F., Moricz, N. and Vancso, J.P. 2014. Reviewing the climate change reviewers: exploring controversy through report references and citations. *Geoforum*, **56**, 17–34.

Johnson, M.F. and Wilby, R.L. 2013. Shield or not to shield: effects of solar radiation on water temperature sensor accuracy. *Water*, **5**, 1622–1637.

Johnson, M.F. and Wilby, R.L. 2015. Seeing the landscape from the trees: metrics to guide riparian shade management in river catchments. *Water Resources Research*, **51**, 3754–3767.

Johnson, M.F., Wilby, R.L. and Toone, J.A. 2014. Inferring air–water temperature relationships from river and catchment properties. *Hydrological Processes*, **28**, 2912–2928.

Johnstone, S. and Mazo, J. 2011. Global warming and the Arab spring. *Survival: Global Politics and Strategy*, **53**, 11–17.

Jones, L., Dougill, A., Jones, R.G., et al. 2015. Ensuring climate information guides long-term development. *Nature Climate Change*, **5**, 812–814.

Jones, P.D. 2008. Historical climatology: a state of the art. *Weather*, **63**, 181–186.

Jones, P.D. and Mann, M.E. 2004. Climate over past millennia. *Review of Geophysics*, **42**, RG2002.

Jones, P.D. and Wigley, T.M.L. 2010. Estimation of global temperature trends: what's important and what isn't. *Climatic Change*, **100**, 59–69.

Jones, P.D., Lister, D.H. and Li, Q. 2008. Urbanization effects in large-scale temperature records, with an emphasis on China. *Journal of Geophysical Research*, **113**, D16122.

Jones, P.D., Lister, D.H., Osborn, T.J., et al. 2012. Hemispheric and large-scale land surface air temperature variations: an extensive revision and an update to 2010. *Journal of Geophysical Research*, **117**, D05127.

Junk, J., Kouadio, L., Delfosse, P. and El Jarroudi, M. 2016. Effects of regional climate change on brown rust disease in winter wheat. *Climatic Change*, **135**, 439–451.

Kallis, G. and Zografos, C. 2014. Hydro-climatic change, conflict and security. *Climatic Change*, **123**, 69–82.

Kanamitsu, M. and DeHaan, L. 2011. The added value index: a new metric to quantify the added value of regional models. *Journal of Geophysical Research: Atmospheres*, **116**, D11106.

Karim, P., Qureshi, A.S., Bahramloo, R., and Molden, D. 2012. Reducing carbon emissions through improved irrigation and groundwater management: a case study from Iran. *Agricultural Water Management*, **108**, 52–60.

Karl, T.R., Arguez, A., Huang, B., et al. 2015. Possible artefacts of data biases in the recent global surface warming hiatus. *Science*, **348**, 1469–1472.

Keith, D.W. 2009. Why capture CO_2 from the atmosphere? *Science*, **325**, 1654–1655.

Keller, K., Hall, M., Rim, S.-R., Bradford, D.F. and Oppenheimer, M. 2005. Avoiding dangerous anthropogenic interference with the climate system. *Climatic Change*, **73**, 227–238.

Kellstedt, P.M., Zahran, S. and Vedlitz, A. 2008. Personal efficacy, the information environment, and attitudes towards global warming and climate change in the United States. *Risk Analysis*, **28**, 113–126.

Kelly, N.M.S. 2014. The scientific and political legacy of the UK Climate Projections (UKCP09). *Area*, **46**, 111–113.

Kelly, P.M. and Adger, W.N. 2000. Theory and practice in assessing vulnerability to climate change and facilitating adaptation. *Climatic Change*, **47**, 325–352.

Kendon, E.J., Roberts, N.M., Fowler, H.J., et al. 2014. Heavier summer downpours with climate change revealed by weather forecast resolution model. *Nature Climate Change*, **4**, 570–576.

Kennedy, J., Steinberger, J., Barrie, G., et al. 2009. Greenhouse gas emissions from global cities. *Environmental Science and Technology*, **43**, 7297–7302.

Kern, F., Gaede, J., Meadowcroft, J. and Watson, J. 2016. The political economy of carbon capture and storage: an analysis of two demonstration projects. *Technological Forecasting and Social Change*, **102**, 250–260.

Kerr, R.A. 2009. Amid worrisome signs of warming, 'climate fatigue' sets in. *Science*, **326**, 926–928.

Kerr, R.A. 2013. Forecasting regional climate change flunks its first test. *Science*, **239**, 638.

Kharin, V.V., Zwiers, F.W., Zhang, X. and Wehner, M. 2013. Changes in temperature and precipitation extremes in the CMIP5 ensemble. *Climatic Change*, **119**, 345–357.

Kim, J.J. and Guha-Sapir, D. 2012. Famines in Africa: is early warning early enough? *Global Health Action*, **5**, 18481.

Kintisch, E. 2014. Is Atlantic holding Earth's missing heat? *Science*, **345**, 860–861.

Kniveton, D.R., Smith, C.D. and Black, R. 2012. Emerging migration flows in a changing climate in dryland Africa. *Nature Climate Change*, **2**, 444–447.

Knutti, R. 2008. Why are climate models reproducing the observed global surface warming so well? *Geophysical Research Letters*, **35**, L18704.

Knutti, R. 2010. The end of model democracy? An editorial comment. *Climatic Change*, **102**, 395–404.

Knutti, R. and Sedláček, J. 2012. Robustness and uncertainties in the new CMIP5 climate model projections. *Nature Climate Change*, **3**, 369–373.

Knutti, R., Rogelj, J., Sedláček, J. and Fischer, E.M. 2016. A scientific critique of the two-degree climate change target. *Nature Geoscience*, **9**, 13–18.

Koo, C., Hong, T., Park, H.S. and Yun, G. 2014. Framework for the analysis of the potential of the rooftop photovoltaic system to achieve the net-zero energy solar buildings. *Progress in Photovoltaics*, **22**, 462–478.

Kosaka, Y. and Xie, S.-P. 2013. Recent global-warming hiatus tied to equatorial Pacific surface cooling. *Nature*, **501**, 403–407.

Kossin, J.P., Olander, T.L. and Knapp, K.P. 2013. Trend analysis with a new global record of tropical cyclone intensity. *Journal of Climate*, **26**, 9960–9976.

Koteyko, N., Jaspal, R. and Nerlich, B. 2013. Climate change and 'climategate' in online reader comments: a mixed methods study. *Geographical Journal*, **179**, 74–86.

KPMG, 2014. *Future State 2030: The Global Megatrends Shaping Governments*. www.kpmg.com/ID/en/IssuesAndInsights/ArticlesPublications/Documents/Future-State-2030.pdf [accessed 23/06/16].

Kuhlbrodt, T. and Gregory, J. 2012. Ocean heat uptake and its consequences for the magnitude of sea level rise and climate change. *Geophysical Research Letters*, **39**, L18608.

Kumar, K.K., Kumar, K.R., Ashrit, R.G., et al. 2004. Climate impacts on Indian agriculture. *International Journal of Climatology*, **24**, 1375–1393.

Kundzewicz, Z.W. and Stakhiv, E.Z. 2010. Are climate models "ready for prime time" in water resources management applications, or is more research needed? *Hydrological Sciences Journal*, **55**, 1085–1089.

Kuntzmann, L.E. and Drake, J.L. 2016. Framing matters in communicating crisis. In Drake, J.L., Kontar, Y.Y., Eichelberger, J.C., Rupp, T.S. and Taylor, K.M. (eds.), *Communicating Climate-Change and Natural Hazard Risk and Cultivating Resilience*. Springer, London.

Kure, S., Jang, S., Ohara, N., Kavvas, M.L. and Chen, Z.Q. 2013. Hydrologic impact of regional climate change for the snowfed and glacierfed river basins in the Republic of Tajikistan: hydrological response of flow to climate change. *Hydrological Processes*, **27**, 4057–4070.

Lagmay, A.M.F., Agaton, R.P., Bahala, M.A.C., et al. 2015. Devastating storm surges of Typhoon Haiyan. *International Journal of Disaster Risk Reduction*, **11**, 1–12.

Lahsen, M. 2013. Climategate: the role of the social sciences. *Climatic Change*, **119**, 547–558.

Lancet Commissions, 2015. Health and climate change: policy responses to protect public health. *The Lancet*, **386**, 1861–1914.

Le Quéré, C., Raupach, M.R., Canadell, J.G., et al. 2009. Trends in the sources and sinks of carbon dioxide. *Nature Geoscience*, **2**, 831–836.

Lee, J-W. and Hong, S-Y. 2014. Potential for added value to downscaled climate extremes over Korea by increased resolution of a regional climate model. *Theoretical and Applied Climatology*, **117**, 667–677.

Leichenko, R., Thomas, A. and Barnes, M. 2010. Vulnerability and adaptation to climate change. In Lever-Tracy, C. (ed.), *Routledge Handbook of Climate Change and Society*, Chapter 7. Routledge, Abingdon.

Leiserowitz, A.A., Maibach, E.W., Roser-Renouf, C., Smith, N. and Dawson, E. 2013. Climategate, public opinion, and the loss of trust. *American Behavioural Scientist*, **57**, 818–837.

Lemonsu, A., Beaulant, A.L., Somot, S. and Masson, V. 2014. Evolution of heat wave occurrence over the Paris basin (France) in the 21st century. *Climate Research*, **61**, 75–91.

Lempert, R.J. and Collins, M.T. 2007. Managing the risk of uncertain threshold responses: comparison of robust, optimum and precautionary approaches. *Risk Analysis*, **27**, 1009–1026.

Lenton, T.M. 1998. Gaia and natural selection. *Nature*, **394**, 439–447.

Lenton, T.M., Held, H., Kriegler, E., et al. 2008. Tipping elements in the Earth's climate system. *Proceedings of the National Academy of Sciences of the United States of America*, **105**, 1786–1793.

Lever-Tracy, C. 2010. *Routledge Handbook of Climate Change and Society*. Routledge, Abingdon.

Leviticus, S., Antonov, J.I., Boyer, T.P., et al. 2012. World ocean heat content and thermosteric sea level change (0–2000 m), 1955–2010. *Geophysical Research Letters*, **39**, L10603.

Lewandowsky, S., Ecker, U.K.H., Seifert, C.M. and Schwartz, N. 2012. Misinformation and its correction: continued influence and successful debiasing. *Psychological Science in the Public Interest*, **13**, 106–131.

Lewandowsky, S., Gignac, G.E. and Vaughan, S. 2013. The pivotal role of perceived scientific consensus in acceptance of science. *Nature Climate Change*, **3**, 399–404.

Lewandowsky, S., Oreskes, N., Risbey, J.S., Newell, B.R. and Smithson, M. 2015. Seepage: climate change denial and its effect on the scientific community. *Global Environmental Change*, **33**, 1–13.

Lewandowsky, S., Risbey, J. and Oreskes, N. 2016. The "pause" in global warming: turning a routine fluctuation into a problem for science. *Bulletin of the American Meteorological Society*, **97**, 723–733.

Lewis, N. and Crok, M. 2014. *Oversensitive: How the IPCC Hid the Good News on Global Warming*. Report 12, The Global Warming Policy Foundation, London.

Liepert, B.G. and Lo, F. 2013. CMIP5 update of 'Inter-model variability and biases of the global water cycle in CMIP3 coupled climate models'. *Environmental Research Letters*, **8**, 029401.

Liepert, B.G. and Previdi, M. 2012. Inter-model variability and biases of the global water cycle in CMIP3 coupled climate models. *Environmental Research Letters*, **7**, 014006.

Link, P.M., Brucher, T., Claussen, M., Link, J.S.A. and Scheffran, J. 2015. The nexus of climate change, land use, and conflict: complex human–environment interactions in Northern Africa. *Bulletin of the American Meteorological Society*, **96**, 1561–1564.

Liu, J.G., Mooney, H., Hull, V., et al. 2015b. Systems integration for global sustainability. *Science*, **347**, 1258832.

Liu, M., Zhu, X., Pan, C., et al. 2015a. Spatial variation of near-surface CO_2 concentration during spring in Shanghai. *Atmospheric Pollution Research*, **7**, 31–39.

Liu, Z. 2015. *China's Carbon Emissions Report 2015*. Belfer Center for Science and International Affairs, Harvard Kennedy School, Cambridge, MA.

Liu, Z., He, C., Zhou, Y. and Wu, J. 2014. How much of the world's land has been urbanized, really? A hierarchical framework for avoiding confusion. *Landscape Ecology*, **29**, 763–771.

Lopez, A., Fung, F., New, M., et al. 2009. From climate model ensembles to climate change impacts: a case study of water resource management in the South West of England. *Water Resources Research*, **45**, W08419.

Lovelock, J.E. and Margulis, L. 1974. Atmospheric homeostasis by and for the biosphere: the Gaia hypothesis. *Tellus*, **26**, 2-10.

Lowe, T., Brown, K., Dessai, S., et al. 2006. Does tomorrow ever come? Disaster narrative and public perceptions of climate change. *Public Understanding of Science*, **15**, 435–457.

Lutz, A.F., Immerzeel, W.W., Gobiet, A., Pellicciotti, F. and Bierkens, M.F.P. 2013. Comparison of climate change signals in CMIP3 and CMIP5 multi-model ensembles and implications for Central Asian glaciers. *Hydrology and Earth System Sciences*, **17**, 3661–3677.

MacCracken, M.C. 2009. On the possible use of geoengineering to moderate specific climate change impacts. *Environmental Research Letters*, **4**, 045107.

Maibach, E., Witte, J. and Wilson, K. 2011. "Climategate" undermined belief in global warming among many American TV meteorologists. *Bulletin of the American Meteorological Society*, **92**, 31–37.

Mainka, S.A. and Howard, G.W. 2010. Climate change and invasive species: double jeopardy. *Integrative Zoology*, **5**, 102–111.

Manabe, S. and Wetherald, R.T. 1975. The effects of doubling CO_2 concentration on the climate of a general circulation model. *Journal of Atmospheric Sciences*, **32**, 3–15.

Mann, M. 2014. Earth will cross the climate danger threshold by 2036. *Scientific American*, 1 April 2014.

Maraun, D., Wetterhall, F., Ireson, A.M., et al. 2010. Precipitation downscaling under climate change: recent development to bridge the gap between dynamical models and the end user. *Reviews of Geophysics*, **48**, RG3003.

Markowitz, D.M. and Hancock, J.T. 2015. Linguistic obfuscation in fraudulent science. *Journal of Language and Social Psychology*, **4**, 435–445.

Marotzke, J. and Forster, P.M. 2015. Forcing, feedback and internal variability in global temperature trends. *Nature*, **517**, 565–570.

Martin, G.M. and The HadGEM2 Development Team, 2011. The HadGEM2 family of Met Office Unified Model climate configurations. *Geoscientific Model Development*, **4**, 723–757.

Masato, G., Hoskins, B.J. and Woollings, T. 2013. Winter and summer Northern Hemisphere blocking in CMIP5 models. *Journal of Climate*, **26**, 7044–7059.

Masson, D. and Knutti, R. 2011. Climate model genealogy. *Geophysical Research Letters*, **38**, L08703.

Matthews, T., Mullan, D., Wilby, R.L., Broderick, C. and Murphy, C. 2016. Past and future climate change in the context of memorable seasonal extremes. *Climate Risk Management*, **11**, 37–52.

Maussion, F., Scherer, D., Molg, T., et al. 2014. Precipitation seasonality and variability over the Tibetan Plateau as resolved by the High Asia Reanalysis. *Journal of Climate*, **27**, 1910–1927.

McCright, A.M. 2010. Anti-reflexivity: the American conservative movement's success in undermining climate science and policy. *Theory, Culture and Society*, **27**, 100–133.

McCright, A.M. and Dunlap, R.E. 2003. Defeating Kyoto: the conservative movement's impact on U.S. climate change policy. *Social Problems*, **50**, 348–373.

McCusker, K.E., Battisti, D.S. and Bitz, C.M. 2015. Inability of stratospheric sulfate aerosol injections to preserve the West Antarctic Ice Sheet. *Geophysical Research Letters*, **42**, 4989–4997.

McDonald, S., Oates, C.J., Thyne, M., Timmis, A.J. and Carlile, C. 2015. Flying in the face of environmental concern: why green consumers continue to fly. *Journal of Marketing Management*, **31**, 1503–1528.

McGuffie, K. and Henderson-Sellers, A. 2001. Forty years of numerical climate modelling. *International Journal of Climatology*, **21**, 1067–1109.

McGuffie, K. and Henderson-Sellers, A. 2014. *The Climate Model Primer*, Fourth Edition. John Wiley & Sons, Chichester.

McMichael, A.J. 2012. Insights from past millennia into climatic impacts on human health and survival. *Proceedings of the National Academy of Sciences of the United States of America*, **109**, 4730–4737.

McMichael, A.J., Wilkinson, P., Kovats, R.S., et al. 2008. International study of temperature, heat and urban mortality: the 'ISOTHURM' project. *International Journal of Epidemiology*, **37**, 1121–1131.

Meckler, A.N., Sigman, D.M., Gibson, K.A., et al. 2013. Deglacial pulses of deep-ocean silicate into the subtropical North Atlantic Ocean. *Nature*, **495**, 495–498.

Meinshausen, M., Meinshausen, N., Hare, W., et al. 2009. Greenhouse-gas emission targets for limiting global warming to 2 °C. *Nature*, **458**, 1158–1163.

Mekonnen, M.M. and Hoekstra, A.Y. 2010. A global and high-resolution assessment of the green, blue and grey water footprint of wheat. *Hydrology and Earth System Sciences*, **14**, 1259–1276.

Mendelsohn, R., Emanuel, K., Chonabayashi, S. and Bakkensen, L. 2012. The impact of climate change on tropical cyclone damage. *Nature Climate Change*, **2**, 205–209.

Milfont, T.L., Evans, L., Sibley, C.G., Ries, J. and Cunningham, A. 2014. Proximity to coast is linked to climate change belief. *PLOS ONE*, **9**, e103180.

Millner, A., Calel, R., Stainforth, D.A. and MacKerron, G. 2013. Do probabilistic expert elicitations capture scientists' uncertainty about climate change? *Climatic Change*, **116**, 427–436.

Milly, P.C.D., Betancourt, J., Falkenmark, M., et al. 2008. Stationarity is dead: whither water management? *Science*, **319**, 573–574.

Minx, J., Baiocchi, G., Wiedmann, T., et al. 2013. Carbon footprints of cities and other human settlements in the UK. *Environmental Research Letters*, **8**, 035039.

Mitchell, T.D. and Hulme, M. 1999. Predicting regional climate change: living with uncertainty. *Progress in Physical Geography*, **23**, 57–78.

Monastersky, R. 2009. A burden beyond bearing. *Nature*, **458**, 1091–1094.

Moomaw, W., Burgherr, P., Heath, G., et al. 2011. Annex II: Methodology. In *IPCC Special Report on Renewable Energy Sources and Climate Change Mitigation* [Edenhofer, O., Pichs-Madruga, R., Sokona, Y. et al. (eds.)]. Cambridge University Press, Cambridge.

Moore, J., Kissinger, M. and Rees, W.E. 2013. An urban metabolism and ecological footprint assessment of Metro Vancouver. *Journal of Environmental Management*, **124**, 51–61.

Moore, J.C., Grinsted, A., Guo, X.R., et al. 2015. Atlantic hurricane surge response to geoengineering. *Proceedings of the National Academy of Sciences of the United States of America*, **112**, 13794–13799.

Mora, S., Jalludin, M., Mahdi, D.R. and Ortiz, A. 2010. Modelling and quantifying risk in Djibouti. *Proceedings of the Eleventh Congress of the International Association of Engineering Geology and the Environment*, Auckland, 5-10 September 2010.

Morgan, M.G. and Keith, D.W. 1995. Subjective judgments by climate experts. *Environmental Science and Technology*, **29**, A468–A476.

Morton, O. 2009. Great white hope. *Nature*, **458**, 1097–1100.

Moser, S.C. 2010. Communicating climate change: history, challenges, process and future directions. *WIREs Climate Change*, **1**, 31–53.

Moss, R.H. 1995. Avoiding 'dangerous' interference in the climate system: the roles of values, science and policy. *Global Environmental Change*, **5**, 3–6.

Moss, R.H., Edmonds, J.A., Hibbard, K.A., et al. 2010. The next generation of scenarios for climate change research and assessment. *Nature*, **463**, 747–756.

Mulholland, P.J., Best, G.R., Coutant, C.C., et al. 1997. Effects of climate change on freshwater ecosystems of the south-eastern United States and the Gulf Coast of Mexico. *Hydrological Processes*, **11**, 949-70.

Murari, K.K., Ghosh, S., Patwardhan, A., Daly, E. and Salvi, K. 2015. Intensification of future severe heatwaves in India and their effect on heat stress and mortality. *Regional Environmental Change*, **15**, 569–579.

Murphy, J.M., Booth, B.B.B., Collins, M., et al. 2007. A methodology for probabilistic predictions of regional climate change from perturbed physics ensembles. *Philosophical Transactions of the Royal Society A*, **365**, 1993–2028.

Murphy, J.M., Sexton, D.M.H., Jenkins, G.J., et al. 2009. *UK Climate Projections Science Report: Climate Change Projections*. Met Office Hadley Centre, Exeter.

Murtaugh, P.A. and Schlax, M.G. 2009. Reproduction and the carbon legacies of individuals. *Global Environmental Change*, **19**, 14–20.

Mwangi, E., Wetterhall, F., Dutram E., et al. 2014. Forecasting droughts in East Africa. *Hydrology and Earth System Sciences*, **18**, 611–620.

Myers, T.A., Maibach, E.E., Roser-Renouf, C., Akerlof, K. and Leiserowitz, A.A. 2013. The relationship between personal experience and belief in the reality of global warming. *Nature Climate Change*, **3**, 343–347.

Nachmany, M., Frankhauser, S., Davidová, J., et al. 2015. *The 2015 Global Climate Legislation Summary: A Review of Climate Change Legislation in 99 Countries*. Grantham Institute, London.

Nagle, J.C. 2009. Climate exceptionalism. *Environmental Law*, **40**, 53–88.

Nakamura, S. 2009. *Spatial Analysis of Urban Poverty in Manila, Philippines*. Cornell University. https://courses.cit.cornell.edu/crp408/papers/nakamura.pdf [accessed 22/10/2015]

Nakhooda, S., Watson, C. and Schalatek, L. 2013. *The Global Climate Finance Architecture*. Overseas Development Institute, London.

Nakicenovich, N. and Swart, R. 2000. *IPCC Special Report on Emissions Scenarios (SRES)*. Cambridge University Press, Cambridge.

Nerlich, B., Koteyko, N. and Brown, B. 2010. Theory and language of climate change communication. *WIREs Climate Change*, **1**, 97–110.

Neukkom, R., Gergis, J., Karoly, D.J., et al. 2014. Inter-hemispheric temperature variability over the past millennium. *Nature Climate Change*, **4**, 362–367.

Nicholls, R.J., Marinova, N., Lowe, J.A., et al. 2011. Sea-level rise and its possible impacts given a 'beyond 4°C world' in the twenty-first century. *Philosophical Transactions of the Royal Society A*, **369**, 161–181.

Nicholson, S. 2014. The predictability of rainfall over the Greater Horn of Africa. Part I: Prediction of seasonal rainfall. *Journal of Hydrometeorology*, **15**, 1011–1027.

Nordbo, A., Järvi, L., Haapanala, S., Wood, C.R. and Vesala, T. 2012. Fraction of natural area as main predictor of net CO_2 emissions from cities. *Geophysical Research Letters*, **39**, L20802.

Nordhaus, W.D. 2007. A review of the *Stern Review on the Economics of Climate Change*. *Journal of Economic Literature*, **45**, 686–702.

Normile, D. 2008. China's living laboratory in urbanization. *Science*, **319**, 740–743.

O'Brien, K., Leichenko, R., Kelkar, U., et al. 2004. Mapping vulnerability to multiple stressors: climate change and globalization in India. *Global Environmental Change*, **14**, 303–313.

O'Hare, G. 2011. Updating our understanding of climate in the North Atlantic: the role of global warming and the Gulf Stream. *Geography*, **96**, 5-15.

O'Hare, G., Sweeney, J. and Wilby, R.L. 2005. Changes in the climate system. *Weather, Climate and Climate Change: Human Perspectives*. Pearson Education, Harlow, UK, pp. 143–175.

O'Neill, S. and Nicholson-Cole, S. 2009. "Fear won't do it": Promoting positive engagement with climate change through visual iconic representations. *Science Communication*, **30**, 355–379.

O'Neill, S.J., Osborn, T.J., Hulme, M., Lorenzoni, I. and Watkinson, A.R. 2008. Using expert knowledge to assess uncertainties in future polar bear populations under climate change. *Journal of Applied Ecology*, **45**, 1649–1659.

Obradovich, N. and Guenther, S.M. 2016. Collective responsibility amplifies mitigation behaviours. *Climatic Change*, **137**, 307–319.

Ogutu, J.O. and Owen-Smith, N. 2003. ENSO, rainfall and temperature influences on extreme population decline among Africa savannah ungulates. *Ecology Letters*, **6**, 412–419.

Olivier, J.G.J., Janssens-Maenhout, G., Muntean, M. and Peters, J.A.H.W. 2015. *Trends in global CO_2 emissions: 2015 Report*. PBL Netherlands Environmental Assessment Agency, and the European Commission Joint Research Centre. The Hague, The Netherlands.

Olson, R.L. 2012. Soft geoengineering: a gentler approach to addressing climate change. *Environment: Science and Policy for Sustainable Development*, **54**, 29–39.

Opiyo, E.O., Wasonga, O.V. and Nyangito, M.M. 2014. Measuring household vulnerability to climate-induced stresses in pastoral rangelands of Kenya: implications for resilience programming. *Pastoralism*, **4**, 10.

Oppenheimer, M. and Petsonk, A. 2005. Article 2 of the UNFCCC: historical origins, recent interpretations. *Climatic Change*, **73**, 195–226.

Oppenheimer, M., Little, C.M. and Cooke, R.M. 2016. Expert judgment and uncertainty quantification for climate change. *Nature Climate Change*, **6**, 445–451.

Oreskes, N. 2004. The scientific consensus on climate change. *Science*, **306**, 1686.

Orlove, B., Lazrus, H., Hovelsrud, G.K. and Giannini, A. 2014. Recognition and responsibilities on the origins and consequences of the uneven attention to climate change around the world. *Current Anthropology*, **55**, 249–275.

Otto, F.E.L., Massey, N., van Oldenborgh, G. J., Jones, R. G. and Allen, M.R. 2012. Reconciling two approaches to attribution of the 2010 Russian heat wave. *Geophysical Research Letters*, **39**, L04702.

Overseas Development Institute (ODI), 2015. *Geography of Poverty, Climate and Disasters II*. Overseas Development Institute, London.

Oxfam, 2010. *Climate Change, Shrinking Glaciers and Poverty in Tajikistan*. Oxfam International, Dushanbe, Tajikistan.

Pacsi, A.P., Alhajeri, N.S., Webster, M.D., Webber, M.E. and Allen, D.T. 2013. Changing the spatial location of electricity generation to increase water availability in areas with drought: a feasibility study and quantification of air quality impacts in Texas. *Environmental Research Letters*, **8**, 035029.

Paeth, H., Scholten, A., Friederichs, P. and Hense, A. 2008. Uncertainties in climate change prediction: El-Niño–Southern Oscillation and monsoons. *Global and Planetary Change*, **60**, 265–288.

PAGES 2k Consortium, 2013. Continental-scale temperature variability during the past two millennia. *Nature Geoscience*, **6**, 339–346.

Pandey, D., Agrawal, M. and Pandey, J.S. 2011. Carbon footprint: current methods of estimation. *Environmental Monitoring and Assessment*, **178**, 135–160.

Parker, D.E. 2004. Large-scale warming is not urban. *Nature*, **432**, 290.

Parker, D.E. 2010. Urban heat island effects on estimates of observed climate change. *WIREs Climate Change*, **1**, 123–133.

Parry, M., Lowe, J. and Hanson, C. 2009. Overshoot, adapt and recover. *Nature*, **485**, 1102–1103.

Parson, E.A. and Keith, D.W. 2013. End the deadlock on governance of geoengineering research. *Science*, **239**, 1278–1279.

Pearce, W., Brown, B., Nerlich, B. and Koteyko, N. 2015. Communicating climate change: conduits, content, and consensus. *WIREs Climate Change*, **6**, 613–626.

Pfaff, A., Broad, K. and Glantz, M. 1999. Who benefits from climate forecasts? *Nature*, **397**, 646.

Pidgeon, N. and Fischhoff, B. 2011. The role of social and decision sciences in communicating uncertain climate risks. *Nature Climate Change*, **1**, 35–41.

Pielke, R.A. Jr. 2009. The British Climate Change Act: a critical evaluation and proposed alternative approach. *Environmental Research Letters*, **4**, 1–7.

Pielke, R.A. Jr., Wigley, T.M.L. and Green, C. 2008. Dangerous assumptions: how big is the energy challenge of climate change? *Nature*, **452**, 531–532.

Pielke, R.A. Sr. 2003. Heat storage within the Earth system. *Bulletin of the American Meteorological Society*, **84**, 331–335.

Pielke, R.A. Sr. and Wilby, R.L. 2012. Regional climate downscaling: what's the point? *Eos*, **93**, 52–53.

Pielke, R.A. Sr., Beven, K., Brasseur, G., et al. 2009. Climate change: the need to consider human forcings beside greenhouse gases. *Eos*, **90**, 413.

Pierce, D.W., Cayan, D.R., Maurer, E.P., Abatzoglou, J.T. and Hegewisch, K.C. 2015. Improved bias correction techniques for hydrological simulations of climate change. *Journal of Hydrometeorology*, **16**, 2421–2442.

Piguet, E. 2010. Linking climate change, environmental degradation, and migration: a methodological overview. *WIREs Climate Change*, **1**, 517–524.

Pirtle, Z., Meyer, R. and Hamilton, A. 2010. What does it mean when climate models agree? A case for assessing independence among general circulation models. *Environmental Science and Policy*, **13**, 351–361.

Plutzer, E., McCaffrey, M., Hannah, A.L., et al. 2016. Climate confusion among U.S. teachers. *Science*, **351**, 664–666.

Poff, N.L., Brown, C.M., Grantham, T.E., et al. 2016. Sustainable water management under future uncertainty with eco-engineering decision scaling. *Nature Climate Change*, **6**, 25–34.

Poortinga, W., Spence, A., Whitmarsh, L., Capstick, S. and Pidgeon, N.F. 2011. Uncertain climate: an investigation into public scepticism about anthropogenic climate change. *Global Environmental Change*, **21**, 1015–1024.

Porio, E. 2011. Vulnerability, adaptation, and resilience to floods and climate change-related risks among marginal, riverine communities in Metro Manila. *Asian Journal of Social Science*, **39**, 425–445.

Porio, E. 2014. Climate change vulnerability and adaptation in Metro Manila challenging governance and human security of urban poor communities. *Asian Journal of Social Science*, **42**, 75-102.

Prein, A.F., Gobiet, A., Suklitsch, M., et al. 2013. Added value of convective permitting seasonal simulations. *Climate Dynamics*, **41**, 2655–2677.

President's Science Advisory Committee, 1965. *Restoring the Quality of Our Environment: Report of the Environmental Pollution Panel*. The White House, Washington DC. http://dge.stanford.edu/labs/caldeiralab/Caldeira%20downloads/PSAC,%201965,%20Restoring%20the%20Quality%20of%20Our%20Environment.pdf [accessed 21/1/16]

Prudhomme, C., Giuntoli, I., Robinson, E.L., et al. 2014. Hydrological droughts in the 21st century, hotspots and uncertainties from a global multi model ensemble experiment. *Proceedings of the National Academy of Sciences of the United States of America*, **111**, 3262–3267.

Prudhomme, C., Wilby, R.L., Crooks, S., Kay, A.L. and Reynard, N.S. 2010. Scenario-neutral approach to climate change impact studies: application to flood risk. *Journal of Hydrology*, **390**, 198–209.

Purse, B.V., Mellor, P.S., Rogers, D.J., et al. 2005. Climate change and the recent emergence of bluetongue in Europe. *Nature Reviews Microbiology*, **3**, 171–181.

Racherla, P.N., Schindell, D.T. and Faluvegi, G.S. 2012. The added value to global model projections of climate change by dynamical downscaling: a case study over the continental US using the GISS-ModelIE2 and WRF models. *Journal of Geophysical Research: Atmospheres*, **117**, D20118.

Rahmstorf, S. and Coumou, D. 2011. Increase of extreme events in a warming world. *Proceedings of the National Academy of Sciences of the United States of America*, **108**, 17905–17909.

Raible, C.C., Brönnimann, S., Auchmann, R., et al. 2016. Tambora 1815 as a test case for high impact volcanic eruptions: Earth system effects. *WIREs Climate Change*, **7**, 569–589.

Rajaratnam, B., Romano, J., Tsiang, M. and Diffenbaugh, N.S. 2015. Debunking the climate hiatus. *Climatic Change*, **133**, 129–140.

Randall, D.A., Wood, R.A., Bony, S., et al. 2007. Climate models and their evaluation. In *Climate Change 2007: The Physical Science Basis. Contribution of Working Group I to the Fourth Assessment Report of the Intergovernmental Panel on Climate Change* [Solomon, S., Qin, D., Manning, M., et al. (eds.)]. Cambridge University Press, Cambridge.

Ranger, N., Reeder, T. and Lowe, J. 2013. Addressing 'deep' uncertainty over long-term climate in major infrastructure projects: four innovations of the Thames Estuary 2100 Project. *EURO Journal on Decision Processes*, **1**, 233–262.

Rasul, G. 2014. Food, water, and energy security in South Asia: a nexus perspective from the Hindu Kush Himalayan region. *Environmental Science and Policy*, **39**, 35–48.

Rasul, G. and Sharma, B. 2015. The nexus approach to water–energy–food security: an option for adaptation to climate change. *Climate Policy*, **16**, 682–702.

Rayner, S., Redgwell C., Savulescu, J., Pidgeon, N. and Kruger, T. 2009. *Memorandum on Draft Principles for the Conduct of Geoengineering Research.* House of Commons Science and Technology Committee, The Stationery Office, London.

Reiche, D. 2010. Renewable energy policies in Gulf countries: a case study of the carbon-neutral "Masdar City" in Abu Dhabi. *Energy Policy*, **38**, 378–382.

Reichler, T. and Kim, J. 2008. How well do coupled models simulate today's climate? *Bulletin of the American Meteorological Society*, **89**, 303–311.

Revich, B. and Shaposhnikov, D. 2008a. Temperature-induced excess mortality in Moscow, Russia. *International Journal of Biometeorology*, **52**, 367–374.

Revich, B. and Shaposhnikov, D. 2008b. Excess mortality during heat waves and cold spells in Moscow, Russia. *Occupational and Environmental Medicine*, **65**, 691–696.

Riahi, K., Grubler, A. and Nakicenovic, N. 2007. Scenarios of long-term socio-economic and environmental development under climate stabilization. *Technological Forecasting and Social Change*, **74**, 887–935.

Rial, J.A., Pielke Sr, R.A., Beniston, M., et al. 2004. Nonlinearities, feedbacks and critical thresholds within the Earth's climate system. *Climatic Change*, **65**, 11–38.

Ridley, D.A., Solomon, S., Barnes, J.E., et al. 2014. Total volcanic stratospheric aerosol optical depths and implications for global climate change. *Geophysical Research Letters*, **41**, 7763–7769.

Risbey, J.S. 2008. The new climate discourse: alarmist or alarming? *Global Environmental Change*, **18**, 26–37.

Robock, A. 2008. Twenty reasons why geoengineering may be a bad idea. *Bulletin of the Atomic Scientists*, **64**, 14–18.

Rockström, J., Lannerstad, M. and Falkenmark, M. 2007. Assessing the water challenge of a new green revolution in developing countries. *Proceedings of the National Academy of Sciences of the United States of America*, **104**, 6253–6260.

Rogelj, J., Meinshaussen, M., Sedláček, J. and Knutti, R. 2014. Implications of potentially lower climate sensitivity on climate projections and policy. *Environmental Research Letters*, **9**, 031003.

Rosenau, J. 2012. Science denial: a guide for scientists. *Trends in Microbiology*, **20**, 567–569.

Rothausen, S.G.S.A. and Conway, D. 2011. Greenhouse gas emissions from energy use in the water sector. *Nature Climate Change*, **1**, 210–219.

Royal Commission on Environmental Pollution (RCEP), 2011. *Demographic Change and the Environment*. Her Majesty's Stationery Office, London.

Royal Society, 2009. *Geoengineering the Climate: Science, Governance and Uncertainty*, Policy Document 10/09. Royal Society, London.

Rufat, S., Tate, E., Burton, C.G. and Maroof, A.S. 2015. Social vulnerability to floods: review of case studies and implications for measurement. *International Journal of Disaster Risk Reduction*, **14**, 470–486.

Sabine, C.L., Feely, R.A., Gruber, N., et al. 2004. The oceanic sink for anthropogenic CO_2. *Science*, **305**, 367–371.

Salazar, J. and Meil, J. 2009. Prospects for carbon-neutral housing: the influence of greater wood use on the carbon footprint of a single-family residence. *Journal of Cleaner Production*, **17**, 1563–1571.

Salehyan, I. 2008. From climate change to conflict? No consensus yet. *Journal of Peace Research*, **45**, 315–326.

Sampei, Y. and Aoyagi-Usui, M. 2009. Mass-media coverage, its influence on public awareness of climate-change issues, and implications for Japan's national campaign to reduce greenhouse gas emissions. *Global Environmental Change*, **19**, 203–212.

Samy, A.M. and Peterson, A.T. 2016. Climate change influences on the global potential distribution of bluetongue virus. *PLOS ONE*, **11**, e0150489.

Sanderson, M. 2010. *Changes in the Frequency of Extreme Rainfall Events for Selected Towns and Cities*. Met Office, Exeter, UK.

Sarewitz, D. 2011. Does climate change knowledge really matter? *WIREs Climate Change*, **2**, 475–481.

Save the Children & Oxfam, 2012. *A Dangerous Delay*. Oxfam GB, Oxford.

Scaife, A.A., Kucharski, F., Folland, C.K., et al. 2009. The CLIVAR C20C project: selected twentieth century climate events. *Climate Dynamics*, **33**, 603–614.

Schäfer, M.S. 2012. Online communication on climate change and climate politics: a literature review. *WIREs Climate Change*, **3**, 527–543.

Schellnhuber, H.J., Cramer, W., Nakicenovic, N., Wigley, T. and Yohe, G. (eds.) 2006. *Avoiding Dangerous Climate Change*. Cambridge University Press, Cambridge.

Schewe, J., Heinke, J., Gerten, D., et al. 2014. Multi-model assessment of water scarcity under climate change. *Proceedings of the National Academy of Sciences of the United States of America*, **111**, 3245–3250.

Schmidli, J., Frei, C. and Vidale, P.L. 2006. Downscaling from GCM precipitation: a benchmark for dynamical and statistical downscaling methods. *International Journal of Climatology*, **26**, 679–689.

Schneider, S.H. 1972. Cloudiness as a global climatic feedback mechanism: the effects on the radiation balance and surface temperature of variations in cloudiness. *Journal of the Atmospheric Sciences*, **29**, 1413–1422.

Schneider, S.H. 1996. Geoengineering: could – or should – we do it? *Climatic Change*, **33**, 291–302.

Scholze, M., Knorr, W., Arnell, N.W. and Prentice, I.C. 2006. A climate-change risk analysis for world ecosystems. *Proceedings of the National Academy of Sciences of the United States of America*, **103**, 13116–13120.

Schuur, E.A.G., Abbott, B.W., Bowden, W.B., et al. 2013. Expert assessment of vulnerability of permafrost carbon to climate change. *Climatic Change*, **119**, 359–374.

Scott, C.A. 2011. The water–energy–climate nexus: resources and policy outlook for aquifers in Mexico. *Water Resources Research*, **47**, W00L04.

Seitz, R. 2011. Bright water: hydrosols, water conservation and climate change. *Climatic Change*, **105**, 365–381.

Seneviratne, S.I., Donat, M.G., Pitman, A.J., Knutti, R. and Wilby, R.L. 2016. Allowable CO_2 emissions based on regional and impact-related climate targets. *Nature*, **529**, 477–483.

Seneviratne, S.I., Wilhelm, M., Stanelle, T., et al. 2013. Impact of soil moisture–climate feedbacks on CMIP5 projections: first result from the GLACE-CMIP5 experiment. *Geophysical Research Letters*, **40**, 5212–5217.

Shahin, M. 2007. *Water Resources and Hydrometeorology of the Arab Region*. Water Science and Technology Library, Springer, Dordrecht, The Netherlands.

Shanahan, M. 2007. *Talking About a Revolution: Climate Change and the Media*. COP13 Briefing and Opinion Papers. International Institute for Environment and Development, London.

Shaposhnikov, D., Revich, B., Bellander, T., et al. 2014. Mortality related to air pollution with Moscow heat wave and wildfire of 2010. *Epidemiology*, **25**, 359–364.

Sheffield, J., Wood, E.F., Chaney, N., et al. 2014. A drought monitoring and forecasting system for sub-Sahara African water resources and food security. *Bulletin of the American Meteorological Society*, **95**, 861–882.

Shepherd, S., Zharkov, S.I. and Zharkov, V.V. 2014. Prediction of solar activity from solar background magnetic field variations in cycles 21–23. *The Astrophysical Journal*, **795**, 46.

Sheridan, S.C. and Allen, M.J. 2015. Changes in the frequency and intensity of extreme temperature events and human health concerns. *Current Climate Change Reports*, **1**, 155–162.

Shukla, J., Palmer, T.N., Hagedorn, R., et al. 2010. Toward a new generation of world climate research and computing facilities. *Bulletin of the American Meteorological Society*, **91**, 1407–1412.

Shukla, S., McNally, A., Husak, G. and Funk, C. 2014. A seasonal agricultural drought forecast system for food-insecure regions of East Africa. *Hydrology and Earth System Sciences*, **18**, 3907–3921.

Smith, J.B., Schneider, S.H., Oppenheimer, M., et al., 2009. Assessing dangerous climate change through an update of the Intergovernmental Panel on Climate Change (IPCC) "reasons for concern". *Proceedings of the National Academy of Sciences of the United States of America*, **106**, 4133–4137.

Social Weather Station, 2014. *Survey on the Impact of Typhoon Yolanda on Filipino Households*. The Asia Foundation. www.sws.org.ph/ pr20140408.htm [accessed 21/10/2015].

Soden, B.J., Wetherald, R.T., Stenchikov, G.L. and Robock, A. 2002. Global cooling after the eruption of Mount Pinatubo: a test of climate feedback by water vapor. *Science*, **296**, 727–730.

Soegaard, H. 2003. Towards a spatial CO_2 budget of a metropolitan region based on textual image classification and flux measurements. *Remote Sensing of Environment*, **87**, 283–294.

Solomon, S., Plattner, G.-K., Knutti, R. and Friedlingstein, P. 2009. Irreversible climate change due to carbon dioxide emissions. *Proceedings of the National Academy of Sciences of the United States of America*, **106**, 1704–1709.

Spiegelhalter, D. and Riesche, H. 2008. Bacon sandwiches and middle-class drinkers: the risk of communicating risk. *Significance*, **1**, 30–33.

Srinivasan, U.T. 2010. Economics of climate change: risk and responsibility by world region. *Climate Policy*, **10**, 298–316.

Stauning, P. 2014. Reduced solar activity disguises global temperature rise. *Atmospheric and Climate Sciences*, **4**, 60–63.

STDE Consortium, 2007. *Etude d'identification des ouvrages de rétention des ruissellements et de recharge des nappes sur l'ensemble du bassin versant*. Projet d'Aménagement Intégré de L'Oued Ambouli, Volume 1, Contrat Cadre Europeaid/11860/C/SV/multi– Lot N°2. République de Djibouti.

Steffen, W., Richardson, K., Rockström, J., et al. 2015. Planetary boundaries: guiding human development on a changing planet. *Science*, **347**, 736–746.

Stehfest, E., Bouwman, L., van Vuuren, D.P., et al. 2009. Climate benefits of changing diet. *Climatic Change*, **95**, 83–102.

Steinberg, P.E. and Shields, R. (eds.) 2008. *What Is a City? Rethinking the Urban after Hurricane Katrina*. University of Georgia Press, Athens, GA.

Stephens, G.L., L'Ecuyer, T., Forbes, R., et al. 2010. Dreary state of precipitation in global models. *Journal of Geophysical Research*, **115**, D24211.

Stephens, G.L., Li, J., Wild, M., et al. 2012. An update on Earth's energy balance in light of the latest global observations. *Nature Geoscience*, **5**, 691–696.

Stephens, J.C. and Keith, D.W. 2008. Assessing geochemical carbon management. *Climatic Change*, **90**, 217–242.

Stern, N. 2007. *The Economics of Climate Change: The Stern Review.* Cambridge University Press, Cambridge.

Stillwell, A.S., King, C.W., Webber, M.E., Duncan, I.J. and Hardberger, A. 2011. The energy–water nexus in Texas. *Ecology and Society*, **16**, 2.

Stive, M.J.F., Fresco, L.O., Kabat, P., Parmet, B.W.A.H. and Veerman, C.P. 2011. How the Dutch plan to stay dry over the next century. *Proceedings of the Institution of Civil Engineers*, **164**, 114–121.

Stone, B., Hess, J.J. and Frumkin, H. 2010. Urban form and extreme heat events: are sprawling cities more vulnerable to climate change than compact cities? *Environmental Health Perspectives*, **118**, 1425–1428.

Stott, P., Good, P., Jones, G., Gillett, N. and Hawkins, E. 2013. The upper end of climate model temperature projections is inconsistent with past warming. *Environmental Research Letters*, **8**, 014024.

Stott, P., Stone, D.A. and Allen, M.R. 2004. Human contribution to the European heatwave of 2003. *Nature*, **432**, 610–614.

Street, R.B., Steynor, A., Bowyer, P. and Humphrey, K. 2009. Delivering and using the UK climate projections 2009. *Weather*, 64, 227–231.

Suarez, P. and Mendler de Suarez, J. 2012. *Paying for Predictions: Original Game Rules and Guidance for Facilitators.* Red Cross/Red Crescent. www.climatecentre.org/resources-games/paying-for-predictions.

Suckall, N., Fraser, E. and Forster, P. 2016. Reduced migration under climate change: evidence from Malawi using an aspirations and capabilities framework. *Climate and Development*, doi:10.1080/17565529.2016.1149441.

Sugar, L., Kennedy, C. and Leman, E. 2012. Greenhouse gas emissions from Chinese cities. *Journal of Industrial Ecology*, **16**, 552–563.

Sultan, B., Labadi, K., Guegan, J. and Janicot, S. 2005. Climate drives the meningitis epidemic onset in West Africa. *PLOS Medicine*, **2**, 43–49.

Svanes, E. and Aronsson, A.K.S. 2013. Carbon footprint of a Cavendish banana supply chain. *The International Journal of Life Cycle Assessment*, **18**, 1450–1464.

Szerszynski, B., Kearnes, M., Macnaghten, P., Owen, R. and Stilgoe, J. 2013. Why solar radiation management geoengineering and democracy won't mix. *Environment and Planning A*, **45**, 2809–2816.

Tablazon, J., Caro, C.V., Lagmay, A.M.F., et al. 2015. Probabilistic storm surge inundation maps for Metro Manila based on Philippine public storm warning signals. *Natural Hazards and Earth System Sciences*, **15**, 557–570.

Takayabu, I., Hibino, K., Sasaki, H., et al. 2015. Climate change effects on the worst-case storm surge: a case study of Typhoon Haiyan. *Environmental Research Letters*, **10**, 064011.

Tang, S. and Dessai, S. 2012. Usable science? The UK Climate Projections 2009 and decision support for adaptation planning. *Weather, Climate and Society*, **4**, 300–313.

Taylor, C.M., de Jeu, R.A.M., Guichard, F., Harris, P.P. and Dorigo, W.A. 2012. Afternoon rain more likely over drier soils. *Nature*, **489**, 423–426.

Taylor, K.E. 2001. Summarizing multiple aspects of model performance in a single diagram. *Journal of Geophysical Research*, **106**, 7183–7192.

Thomas, C.D., Cameron, A., Green, R.E., et al. 2004. Extinction risk from climate change. *Nature*, **427**, 145–148.

Thompson, D.W.J. and Wallace, J.M. 2001. Regional climate impacts of the Northern Hemisphere Annular Mode. *Science*, **293**, 85–89.

Thompson, D.W.J., Kennedy, J.J., Wallace, J.M. and Jones, P.D. 2008. A large discontinuity in the mid-twentieth century in observed global-mean surface temperature. *Nature*, **453**, 646–649.

Thomson, A.M., Calvin, K.V., Smith, S.J., et al. 2011. RCP4.5: a pathway for stabilization of radiative forcing by 2100. *Climatic Change*, **109**, 77–94.

Thuiller, W., Broennimann, O., Hughes, G., et al. 2006. Vulnerability of African mammals to anthropogenic climate change under conservative land transformation assumptions. *Global Change Biology*, **12**, 424–440.

Tilman, D., Socolow, R., Foley, J.A., et al. 2009. Beneficial biofuels: the food, energy, and environment trilemma. *Science*, **325**, 270–271.

Tilmes, S., Mills, M.J., Niemeier, U., et al. 2015. A new Geoengineering Model Intercomparison Project (GeoMIP) experiment designed for climate and chemistry models. *Geoscientific Model Development*, **8**, 43–49.

Tisdal, S. 2012. East Africa's drought: the avoidable disaster. *The Guardian*, 18 January 2012. www.guardian.co.uk/world/2012/jan/18/east-africa-drought-disaster-report [accessed 3/2/2015].

Tobler, C., Visschers, V.H.M. and Siegrist, M. 2012. Addressing climate change: determinants of consumers' willingness to act and to support policy measures. *Journal of Environmental Psychology*, **32**, 197–207.

Tokarska, K.B. and Zickfeld, K. 2015. The effectiveness of net negative carbon dioxide emissions in reversing anthropogenic climate change. *Environmental Research Letters*, **10**, 094013.

Tollefson, J. 2014. The case of the missing heat. *Nature*, **505**, 276–278.

Tollefson, J. 2015. The 2 °C dream. *Nature*, **527**, 436–438.

Toone, J.A., Wilby, R.L. and Rice, S. 2011. Surface-water temperature variations and river corridor properties. In *Water Quality: Current Trends and Expected Climate Change Impacts*, IAHS Publication 348. International Association of Hydrological Sciences, Wallingford, UK, pp. 129–134.

Tselioudis, G., Douvis, C. and Zerefis, C. 2012. Does dynamical downscaling introduce novel information in climate model simulations of precipitation change over a complex topography region? *International Journal of Climatology*, **32**, 1572–1578.

Tuck, A.F., Donaldson, D.J., Hitchman, M.H., et al. 2008. On geoengineering with sulphate aerosols in the tropical upper troposphere and lower stratosphere. *Climatic Change*, **90**, 315–331.

United Nations Development Programme (UNDP), 2014. *Sustaining Human Progress: Reducing Vulnerabilities and Building Resilience*. United Nations, New York.

United Nations Framework Convention on Climate Change (UNFCCC), 1992. https://unfccc.int/resource/docs/convkp/conveng.pdf [accessed 21/1/16].

Unruh, G.C. 2000. Understanding carbon lock in. *Energy Policy*, **28**, 817–830.

van der Linden, S.L., Leiserowitz, A.A., Feinberg, G.D. and Mailbach, E.W. 2015. The scientific consensus on climate change as a gateway belief: experimental evidence. *PLOS ONE*, **10**, e0118489.

van Vuuren, D.P., van Soest, H., Riahi, K., et al. 2016. Carbon budgets and energy transition pathways. *Environmental Research Letters*, **11**, 075002.

Vaughan, C. and Dessai, S. 2014. Climate services for society: origins, institutional arrangements, and design elements for an evaluation framework. *WIREs Climate Change*, **5**, 587–603.

Vaughan, D.G. and Spouge, J.R. 2002. Risk estimation of collapse of the West Antarctic Ice Sheet. *Climatic Change*, **52**, 65–91.

Vaughan, N.E. and Lenton, T.M. 2011. A review of climate geoengineering proposals. *Climatic Change*, **109**, 745–790.

Verner, D. 2013. *Tunisia in a Changing Climate: Assessment and Actions for Increased Resilience and Development*. World Bank, Washington DC.

Verosub, K.L. 2010. Climate science in a postmodern world. *Eos*, **91**, 291–292.

Vincent, K., Dougill, A.J., Dixon, J.L., Stringer, L.C. and Cull, T. 2015. Identifying climate service needs for national planning: insights from Malawi. *Climate Policy*, doi:10.1080/14693062.2015.1075374.

Vinkers, C.H., Tijdink, J.K. and Otte, W.M. 2015. Use of positive and negative words in scientific PubMed abstracts between 1974 and 2014: retrospective analysis. *British Medical Journal*, **351**, h6467.

Virgoe, J. 2009. International governance of a possible geoengineering intervention to combat climate change. *Climatic Change*, **95**, 103–119.

von Schuckmann, K., Palmer, M.D., Trenberth, K.E., et al. 2016. An imperative to monitor Earth's energy imbalance. *Nature Climate Change*, **6**, 138–144.

Wade, S.D., Rance, J. and Reynard, N. 2013. The UK Climate Change Risk Assessment 2012: assessing the impacts on water resources to inform policy makers. *Water Resources Management*, **27**, 1085–1109.

WAIS Divide Project Members, 2013. Onset of deglacial warming in West Antarctica driven by local orbital forcing. *Nature*, **500**, 440–444.

Walsh, C.L., Blenkinsop, S., Fowler, H.J., et al. 2016. Adaptation of water resource systems to an uncertain future. *Hydrological Earth System Science*, **20**, 1869–1884.

Walsh, C.L., Dawson, R.J., Hall, J.W., et al. 2011. Assessment of climate change mitigation and adaptation in cities. *Proceedings of the Institution of Civil Engineers: Urban Design and Planning*, **164**, 75–84.

Walther, G.-R., Post, E., Coney, P., et al. 2002. Ecological responses to recent climate change. *Nature*, **416**, 389–395.

Wang, J., Rothausen, S.G.S.A., Conway, D., et al. 2012. China's water–energy nexus: greenhouse-gas emissions from groundwater use for agriculture. *Environmental Research Letters*, **7**, 014035.

Warren, R., de la Nava Santos, S., Arnell, N.W., et al. 2008. Development and illustrative outputs of the Community Integrated Assessment System (CIAS), a multi-institutional modular integrated assessment approach for modelling climate change. *Environmental Modelling and Software*, **23**, 592–610.

Washington, R., Harrison, M., Conway, D., et al. 2006. African climate change: taking the shorter route. *Bulletin of the American Meteorological Society*, **87**, 1355–1366.

Washington, W. 2008. *Odyssey in Climate Modeling, Global Warming, and Advising Five Presidents*, Second Edition. Lulu.com.

Weaver, C.P., Lempert, R.J., Brown, C., et al. 2013. Improving the contribution of climate model information to decision making: the value and demands of robust decision frameworks. *WIREs Climate Change*, **4**, 39–60.

Weber, E.U. and Stern, P.C. 2011. Public understanding of climate change in the United States. *American Psychologist*, **66**, 315–328.

Webster, M., Donohoo, P. and Palmintier, B. 2013. Water–CO_2 trade-offs in electricity generation planning. *Nature Climate Change*, **3**, 1029–1032.

Weigold, M.F. 2001. Communicating science: a review of the literature. *Science Communication*, **23**, 164–193.

Weinthal, E., Zawahri, N. and Sowers, J. 2015. Securitizing water, climate, and migration in Israel, Jordan, and Syria. *International Environmental Agreements: Politics, Law and Economics*, **15**, 293–307.

Weller, E. and Cai, W. 2013. Realism of the Indian Ocean Dipole in CMIP5 models: the implications for climate projections. *Journal of Climate*, **26**, 6649–6659.

Wells, N.C. 2016. The North Atlantic Ocean and climate change in the UK and northern Europe. *Weather*, **71**, 3–6.

Whitehead, P.G., Wilby, R.L., Butterfield, D. and Wade, A.J. 2006. Impacts of climate change on nitrogen in a lowland chalk stream: an appraisal of adaptation strategies. *Science of the Total Environment*, **365**, 260–273.

Whitmarsh, L. 2008. Are flood victims more concerned about climate change than other people? The role of direct experience in risk perception and behavioural response. *Journal of Risk Research*, **11**, 351–374.

Wickham, C., Rohde, R., Muller, R.A., et al. 2013. Influence of urban heating on the global temperature land average using rural sites identified from MODIS classifications. *Geoinformatics and Geostatistics: An Overview*, 1(2).

Wigley, T.M.L. 2006. A combined mitigation/geoengineering approach to climate stabilization. *Science*, **314**, 452–454.

Wilby, R.L. 2007. A review of climate change impacts on the built environment. *Built Environment Journal*, **33**, 31–45.

Wilby, R.L. 2008. Constructing climate change scenarios of urban heat island intensity and air quality. *Environment and Planning B: Planning and Design*, **35**, 902–919.

Wilby, R.L. 2009. *Climate Change Risks and Adaptation Options in Djibouti*. World Bank, Washington, DC.

Wilby, R.L. 2010a. Evaluating climate model outputs for hydrological applications: opinion. *Hydrological Sciences Journal*, **55**, 1090–1093.

Wilby, R.L. 2010b. *Improving the Climate Resilience of Tajikistan's Hydropower Sector: Climate Variability and Change in Tajikistan.* Report on behalf of the European Bank for Reconstruction and Development, London.

Wilby, R.L. 2016. Climate service sector needs robust standards. www.scidev.net/global/climate-change/opinion/climate-service-sector-standards.html [accessed 14/6/16].

Wilby, R.L. and Dawson, C.W. 2013. The Statistical DownScaling Model (SDSM): insights from one decade of application. *International Journal of Climatology*, **33**, 1707–1719.

Wilby, R.L. and Dessai, S. 2010. Robust adaptation to climate change. *Weather*, **65**, 180–185.

Wilby, R.L. and Harris, I. 2006. A framework for assessing uncertainties in climate change impacts: low flow scenarios for the River Thames, UK. *Water Resources Research*, **42**, W02419.

Wilby, R.L. and Vaughan, K. 2011. The hallmarks of organizations that are adapting to climate change. *Water and Environment Journal*, **25**, 271–281.

Wilby, R.L. and Wigley, T.M.L. 1997. Downscaling general circulation model output: a review of methods and limitations. *Progress in Physical Geography*, **21**, 530–548.

Wilby, R.L. and Wigley, T.M.L. 2000. Precipitation predictors for downscaling: observed and general circulation model relationships. *International Journal of Climatology*, **20**, 641–661.

Wilby, R.L. and Yu, D. 2013. Rainfall and temperature estimation for a data sparse region. *Hydrology and Earth System Sciences*, **17**, 3937–3955.

Wilby, R.L., Dawson, C.W., Murphy, C., O'Connor, P. and Hawkins, E. 2014. The Statistical DownScaling Model – Decision Centric (SDSM-DC): conceptual basis and applications. *Climate Research*, **61**, 259–276.

Wilby, R.L., Jones, P.D. and Lister, D. 2011. Decadal variations in the nocturnal heat island of London. *Weather*, **66**, 59–64.

Wilby, R.L., Mora, S., Abdallah, A.O. and Ortiz, S. 2010a. Confronting climate variability and change in Djibouti through risk management. *Proceedings of the Eleventh Congress of the International Association of Engineering Geology and the Environment*, Auckland, 5–10 September 2010.

Wilby, R.L., Orr, H., Watts, G., et al. 2010b. Evidence needed to manage freshwater ecosystems in a changing climate: turning adaptation principles into practice. *Science of the Total Environment*, **408**, 4150–4164.

Wilby, R.L., Troni, J., Biot, Y., et al. 2009. A review of climate risk information for adaptation and development planning. *International Journal of Climatology*, **29**, 1193–1215.

Williams, H.T.P., McMurray, J.R., Kurz, T. and Lambert, F.H. 2015. Network analysis reveals open forums and echo chambers in social media discussions of climate change. *Global Environmental Change*, **32**, 126–138.

Willows, R. and Connell, R. (eds.) 2003. *Climate Adaptation: Risk, Uncertainty and Decision-Making.* UK Climate Impacts Programme Technical Report. UKCIP, Oxford.

Wolf, T. and McGregor, G. 2013. The development of a heatwave vulnerability index for London, United Kingdom. *Weather and Climate Extremes*, **1**, 59–68.

Wolman, A. 1965. The metabolism of cities. *Scientific American*, **213**, 179–190.

Wong, K.V., Paddon, A. and Jimenez, A. 2013. Review of world urban heat islands: Many linked to increased mortality. *Journal of Energy Resources Technology: Transactions of the ASME*, **135**, 022101.

World Bank, 2013. *Turn Down the Heat: Climate Extremes, Regional Impacts, and the Case for Resilience*. World Bank, Washington, DC.

World Health Organisation (WHO), 2014. *Quantitative Risk Assessment of the Effects of Climate Change on Selected Causes of Death, 2030s and 2050s*. WHO, Geneva.

World Meteorological Organisation (WMO), 2008. *Guide to Meteorological Instruments and Methods of Observation*. WMO, Geneva.

World Meteorological Organisation (WMO), 2014. *Atlas of Mortality and Economic Losses from Weather, Climate and Water Extremes (1970–2012)*. WMO, Geneva.

Wright, D., Leigh, R., Kleinberg, J., Abbott, K. and Scheib, J. 2014. New York City can eliminate the carbon footprint of its buildings by 2050. *Energy for Sustainable Development*, **23**, 46–58.

Xia, L., Robock, A., Cole, J., et al. 2014. Solar radiation management impacts on agriculture in China: a case study in the Geoengineering Model Intercomparison Project (GeoMIP). *Journal of Geophysical Research: Atmospheres*, **119**, 8695–8711.

Yates, D., Miller, K.A., Wilby, R.L. and Kaatz, L. 2015. Decision-centric adaptation appraisal for water management across Colorado's Continental Divide. *Climate Risk Management*, **10**, 35–50.

Yuan, X., Wood, E.F., Roundy, J.K. and Pan, M. 2013. CFSv2-based seasonal hydroclimatic forecasts over the conterminous United States. *Journal of Climate*, **26**, 4828–4847.

Zhang, Y., Liu, X., Xiao, R. and Yuan, Z. 2015. Life cycle assessment of cotton T-shirts in China. *The International Journal of Life Cycle Assessment*, **20**, 994–1004.

Zhao, T.T., Horner, M.W. and Sulik, J. 2011. A geographic approach to sectoral carbon inventory: examining the balance between consumption-based emissions and land-use carbon sequestration in Florida. *Annals of the Association of American Geographers*, **101**, 752–763.

Zickfeld, K. and Bruckner, T. 2008. Reducing the risk of Atlantic thermohaline circulation collapse: sensitivity analysis of emissions corridors. *Climatic Change*, **91**, 291–315.

Ziska, L.H., Blumenthal, D.M., Runion, G.B., Hunt Jr, E.R. and Diaz-Soltero, H. 2011. Invasive species and climate change: an agronomic perspective. *Climatic Change*, **105**, 13–42.

Index

Printed in the United States
by Baker & Taylor Publisher Services